D0850621

Living without Oxygen

Living
without
Oxygen

Closed and Open Systems
in Hypoxia Tolerance

Peter W. Hochachka

HARVARD UNIVERSITY PRESS
Cambridge, Massachusetts
London, England
1980

Library of Congress Cataloging in Publication Data

Hochachka, Peter W
 Living without oxygen.

 Includes index.
 1. Oxygen—Physiological effect. 2. Glycolysis.
3. Anaerobiosis. 4. Anoxemia. I. Title.
QP177.H63   591.1′28   79-20221
ISBN 0-674-53670-3

Designed by Mike Fender

To my grandfather,
who taught me how to see nature,
and my father,
who taught me how to understand it

Preface

Although I normally pursue my work in a Faculty of Science and therefore am far removed from medical applications of metabolic biochemistry, the stimulus for writing this book arose during a sabbatical visit to a medical school and a working hospital. There I was introduced to an incredible array of clinical problems that have as their basis some derangement in oxygen-dependent metabolism. There I learned that the two commonest causes of death in our society remain stroke and cardiac arrest. And there I learned that the clinician's view of anaerobic mechanisms is a highly conservative one, often based on frighteningly close encounters with biochemistry of the introductory kind usually during early training. The impression I carried away from numerous grand rounds was that a simple exposition of current thoughts on the biochemical basis of hypoxia and anoxia tolerance might serve a useful and heuristic function.

At that time, the skeletal outline of my book was simple. It called for a broad comparative approach and an overview of anaerobic mechanisms utilized by nature's most capable animal anaerobes. From this exploration, we should be able to perceive what are the most fundamental features of anaerobiosis, shared by all animals, by all organs, tissues, and cells when considered in isolation. Finally, through the use of organisms as an experimental parameter per se, as a window on the world of animals living without oxygen often for astonishing time periods, my intention was to assess what can and what cannot be adjusted in the process of extending the whole organism's tolerance to anoxia or hypoxia. The strategy behind this plan for the book was also simple, for it was my hope that the clarification of adaptable zones in anaerobic metabolism would in turn yield medically useful information. I viewed clarifying what can be modified in anaerobic metabolism as my job;

determination of what may be medically applicable I viewed as the job of perceptive and more clinically directed biochemists.

Whereas these initial thoughts on the book were straightforward, converting them into an essay was not. It required that I course widely through much unfamiliar literature, often spread randomly through biological, biochemical, physiological, or biomedical journals. My reference lists soon became prohibitively long, which is why each chapter lists publications chosen as funnels into each particular subject matter rather than a comprehensive literature in the area.

Also, because of the nature of the material covered, it was important to maintain dialogues with various colleagues around the world. Of these, long discussions with my students, past and present, stand out as being of particular importance to expansion of my own thoughts. Even if it often is difficult to properly ascertain true origins of ideas and interpretations, I feel it is imperative to risk error and pay homage. For chapter 1, my thoughts were most strongly influenced by Daniel E. Atkinson. I leaned heavily upon Tariq Mustafa, Thomas Moon, Ronald Podesta, David Mettrick, and William Hulbert for helping me sort out the literature for chapter 2. Jeremy Fields and Janet Collicutt-Storey contributed more than they probably realize to chapter 3, as did Ab de Zwaan and his colleagues in Holland. Kenneth Storey is my shadow man in chapter 4, while James Ballantyne strongly influenced my thoughts on enzyme regulation in chapter 5. Michael Guppy deserves special thanks for his indirect and direct contributions to chapter 6. Much of the material in chapter 7 has been closely discussed and developed in collaboration with Eric Shoubridge, while earlier aspects of the problem were brought to my attention by F. Ronald Hayes and C. Ladd Prosser. In addition, William Hulbert, Helga Guderley, David Schneider, and Christopher French contributed to my appreciation of anaerobic mechanisms in air-breathing fishes, discussed in chapter 8. As for chapter 9, the classic studies of Per Scholander and Lawrence Irving, published over four decades ago, served as my inlet into metabolic problems during diving in marine mammals. On the practical and experimental side, my inlet into this area was made possible through a collaborative research program with Warren Zapol, Brian Murphy, Michael Snider, Robert Schneider, Mont Liggins, Jesper Qvist, and Robert Creasy, performed in the Antarctic (at McMurdo Station) in 1976 and 1977 and funded by the United States Antarctic Research Program and the National Science and Engineering Research Council (Canada). Warren Zapol was particularly helpful in digesting physiological data for me. By their own contributions, C. R. Taylor and K. Schmidt-Nielsen profoundly influenced my outlook on comparative physiology and stimulated my search for extremists as experimental organisms. Special thanks are also due Dick Taylor for discussing the first draft.

The seed for this book was first planted in my mind during a 1976 expedition into the Amazon, where oxygen is one of the most vicarious of wild environmental parameters faced by aquatic organisms. The kernel of the book took form during a series of lectures (Hypoxia Adaptations in Animals and Man) at the Massachussetts General Hospital, which Warren Zapol encouraged me to prepare. And complete maturation of the project was greatly facilitated by a Fellowship of the Guggenheim Memorial Foundation; some of my work reported in this book also has been supported by N.S.E.R.C. of Canada (with operating, strategic, and negotiated grants).

Finally, despite these intellectual and financial stimuli, none of this would have been possible without the encouragement (we all need when exhausted), the understanding (of why yet another adventure), and the companionship I have received from Brenda, Claire, Gail, Gareth Vasilii, and Laika. In the end, it is they who have supplied the most critical ingredients needed to keep me wondering. And that, after all, is the first step in science.

Vancouver, Canada   P.W.H.

Contents

1   Anaerobic Metabolism: What Can and
        What Cannot Change   1
2   Helminths and the Usefulness of Carbon
        Dioxide   15
3   Coupled Glucose and Amino Acid
        Catabolism in Bivalve Molluscs   27
4   Coupled Glucose and Arginine
        Metabolism in Cephalopods   42
5   Key Elements of Anaerobic Glycolysis   60
6   Integrating Aerobic and Anaerobic
        Glycolysis   79
7   Integrative Mechanisms in
        Hypoxia-Adapted Fish   100
8   Air-Breathing Fish   117
9   Diving Marine Mammals   145
    Epilogue   170
    Index   175

Living
without
Oxygen

Chapter One

Anaerobic Metabolism: What Can and What Cannot Change

Because a limitation of oxygen at the cellular level any-where in the body can lead to serious complications, a great deal of clinical interest has focused on anaerobic metabo-lism in various mammalian organs. All workers have pro-ceeded on at least two fundamental assumptions. First, in order for an organ to sustain significant periods of anoxia, provision must be made for fermentable sources of storage energy, for maintenance of adequate oxidation-reduction (redox) potentials, and for generation of adenosine triphos-phate (ATP). The second assumption is that these funda-mental needs are met by a standard metabolic pathway (an-aerobic glycolysis) for which the fermentable source of energy is glycogen (or glucose), in which redox balance is maintained by a functionally 1:1 coupling of glyceralde-hyde 3-phosphate dehydrogenase to lactate dehydrogenase, and by which adenosine triphosphate is generated in sub-strate-level phosphorylations catalyzed by phosphogly-cerate kinase and pyruvate kinase.

Although these fundamental mechanisms are under-stood by all workers in the field, they are rarely discussed in detail and are usually considered to be highly conservative traits. That is why, in an analysis of organisms on a broad phylogenetic scale to find out what adjustments are or are not possible during hypoxia adaptation some surprises de-velop. Rather than being highly conservative these essential mechanisms in fact appear to supply the raw material that

organisms use to deal with hypoxia, for all three mechanisms are evidently adjustable and closely modulated. Furthermore, although these three mechanisms are essential
for all organisms that are able to sustain some period of anoxia, they are not sufficient for the most effective of anaerobes. The latter develop auxiliary anaerobic mechanisms
that greatly expand their capability to withstand a lack of
oxygen. In many cases these auxiliary mechanisms are
merely exaggerated versions of metabolic processes that are
normally represented in all organisms. Hence, their study in
truly effective anaerobes yields insights into their roles in
"standard" organisms which may be of greater interest to
biomedical science.

Although most of this book will be about animals that
are particularly good at sustaining anoxia or severe hypoxia,
it is useful to begin by expanding on the biochemical mechanisms that supply organisms their raw material for solving
problems of oxygen limitation.

Different Forms of Storage Glycogen

Since the days of Embden and Meyerhof, it has been
evident that carbohydrate, glycogen in particular, is the
major storage form of energy for anaerobic metabolism in
all animals. Interestingly, glycogen can occur in different
forms, and many organisms take advantage of such differences to assist them in sustaining extreme hypoxia. In most
organs, intracellular storage glycogen, a complicated glucose polymer of 1-5 million molecular weight, with 1-4
linkages forming linear sections and 1-6 linkages forming
branching sections, is packaged into small granules 150 to
400 angstroms in diameter. These β-particles are dispersed
throughout the cytoplasm and nuclear regions of the cell.
They do not occur within mitochondria, and in muscle cells
are usually located in an interfibrillar location. When glycogen is abundant in a given organ, these small glycogen granules are usually abundant. At least in skeletal muscle, glycogen granules are associated with enzymes involved in
glycogen synthesis and glycogen degradation. Surprisingly
little is known of the means by which glycogen granules
form during glycogen deposition.

If little is known about the formation of β-particles, even less information is available on the manner of formation of larger α-particles or rosettes. Glycogen rosettes range up to 1000 angstroms in diameter—that is, anywhere from two to five times larger than β-particles—and they seem to represent a more efficient way of packaging glycogen for they are typically found only in cells storing unusually high concentrations of glycogen. Thus they are commonly observed in electron micrographs of liver and kidneys, particularly under certain pathological conditions involving biochemical lesions in glycogen metabolism. Glycogen rosettes are rarely seen in other organs such as the heart, brain, or muscles, but this is not to imply that they cannot be formed here under certain conditions or in certain species. In the heart of the Antarctic Weddell seal, for example, the 1000 angstrom α-particle is a common storage form of glycogen.

Some form of a glycogen body is another specialized way of storing intracellular glycogen. Glycogen bodies vary in shape and detailed structure in different species. In mammalian tissues where they have thus far been studied (ocular muscle, nerve, heart) these appear associated with modified regions of the endoplasmic reticulum (ER), or in muscles, the sarcoplasmic reticulum (SR). Glycogen granules that form in these bodies appear to be about 350 angstroms in diameter and are deposited in very regular arrangement, usually in membrane cisternae.

An unusually rich source of glycogen bodies also is to be found in tuna white muscle, which appears capable of extraordinarily active anaerobic glycolysis; lactate dehydrogenase (LDH), for example, occurs at an activity of nearly 6000 μmol substrate converted to product/min/gm tissue at 25°C. Associated with ample glycolytic machinery is a large amount of glycogen, stored either as β-particles or glycogen bodies. Glycogen bodies appear to be formed primarily in peripheral regions of white muscle fibers and clearly are associated with a complicated membrane system presumed to be specialized sarcolemma (fig. 1.1). In addition, glycogen bodies often seem transferred to interstitial regions, where they are usually found with fully intact, bounding mem-

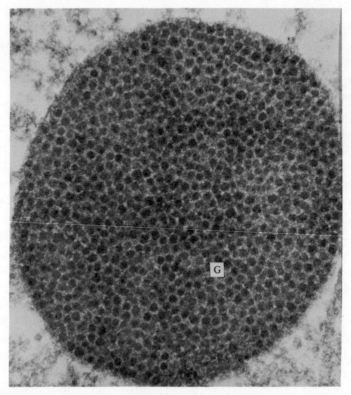

Figure 1.1 A high-magnification (×68,700) electron micrograph of a glycogen body in white muscle of unexercised tuna, showing the regular organization of glycogen granules (G). (From Hulbert et al. 1979.)

branes. Such regions may occupy over 10% of the area of electron micrographs. The matrix of the glycogen bodies appears in electron micrographs as a diffuse material, weakly electron-absorbing. Similar but less numerous glycogen bodies are also found in tuna red muscle and have been reported in heart and red muscle of Amazon air-breathing fishes.

The participation of different glycogen depots in metabolism can be conveniently demonstrated by sampling muscle following bursts of swimming. Micrographs of such samples of tuna white muscle show that glycogen granules can be depleted from all identifiable storage sites, from peripheral and interfibrillar storage pools and from glycogen bodies. However, nothing is known about how these glycogen bodies are formed. How do they interact with their associated membrane components? Are they associated in the usual manner with glycogen phosphorylase or glycogen synthetase? And, since glycogen bodies coexist in the same cells with randomly dispersed β-particles, are they preferentially utilized or preferentially conserved for extreme emergencies? We do not know the answers, but the questions seem to be good ones.

To date, the only other form of deposit glycogen known in vertebrate organisms is termed the glycogen "sea." Glycogen seas have only been observed in highly anaerobic skeletal muscles. They are most prominent in the red muscle of the South American lungfish. Unlike the usual β-particle form of glycogen granules deposited randomly in interfibrillar and peripheral regions of muscle cells, glycogen seas are found sequestered away in specialized peripheral regions of the muscle cells, often, but not always, around the nuclei. Such glycogen seas are formed by a very regular deposition of granules of essentially identical size (about 300 angstroms in diameter). Electron micrographs of such seas show periodic glycogen bodies almost drowned by the abundance of the surrounding granules. How these glycogen seas are formed, again, is unknown. However, something can be said of their probable function. Because they show little or no sign of depletion in muscle cells showing substantial depletion of interfibrillar β-particles, it is probably that glycogen seas serve as a true emergency reservoir and probably are utilized under only the most severe anoxic stress. (See table 1.1 for a summary of the storage forms of glycogen.)

Table 1.1 Storage forms of glycogen.

Form	Size (diameter)	Distribution
Granules (β particles)	Small (150–400Å)	Widely dispersed
Rosettes (α particles)	Large (up to 1000Å)	Interfibrillar, random
Glycogen-membrane associations	Small to large	ER or SR associated
Glycogen "seas"	Small	Patchy

Oxidation-Reduction Potential

In order to sustain anoxia, each oxidative step in fermentative metabolism must be balanced by a reductive step. There are two important aspects to redox regulation that must be considered at the outset: the oxidation state of the cytosol and of the mitochondria must be maintained; and the oxidation state of soluble coenzymes such as nicotinamide adenine dinucleotide (NAD^+/NADH) as well as of bound coenzymes such as flavin adenine dinucleotide (FAD^+/FADH) must be maintained.

Cytosolic Redox Regulation

In the case of the cytosol, the oxidative state is always maintained by an oxidized substrate serving as a hydrogen acceptor. This standard mechanism in mammalian organs is the LDH coupled oxidation-reduction reaction:

$$\text{pyruvate} + \text{NADH} + \text{H}^+ \rightleftharpoons \text{lactate} + \text{NAD}^+$$

Auxiliary mechanisms of redox regulation in the cytosol may involve malate dehydrogenase or α-glycerophosphate dehydrogenase (α-GPDH). In invertebrate tissues, LDH may be absent, and LDH function may be taken over by analogous dehydrogenases (table 1.2). In cephalopods, octopine dehydrogenase serves this function. In bivalve molluscs, alanopine dehydrogenase or strombine dehydrogenase may play this role. Chemically, these enzymes catalyze reductive condensation reactions which are analogous to gluta-

Table 1.2 Standard mechanisms of cytosolic redox regulation.[a]

Enzyme	Substrates	Product
Lactate dehydrogenase	Pyruvate	Lactate
Octopine dehydrogenase	Pyruvate, arginine	Octopine
Alanopine dehydrogenase	Pyruvate, alanine	Alanopine
Strombine dehydrogenase	Pyruvate, glycine	Strombine
Oxtopinate dehydrogenase	Pyruvate, ornithine	Oxtopinate
Lysopine dehydrogenase	Pyruvate, lysine	Lysopine
Histopine dehydrogenase	Pyruvate, histidine	Histopine

[a] Auxiliary mechanisms may involve malate dehydrogenase, α-glycerophosphate dehydrogenase, or glutamate dehydrogenase.

mate dehydrogenase and saccharopine dehydrogenase in mammalian tissues. They are also similar to condensations between pyruvate + ammonia, pyruvate + ornithine, pyruvate + lysine, and pyruvate + histidine—reactions that are known in microbial metabolism. Metabolically, however, these novel dehydrogenases are analogous to LDH. They serve to oxidize glycolytically-formed NADH, generating unique anaerobic end products in consequence: octopine in cephalopods, alanopine or strombine in bivalves. Such enzymes have not been looked for in vertebrate tissues, and a systematic search for their presence is an urgent need. But whatever the outcome of that search, the occurrence of any one of such novel dehydrogenases at significant levels would profoundly alter our perception only of cytosolic redox regulation in anoxic stress. As these are all cytosolic enzymes, their function would not bear on mitochondrial redox regulation.

Mitochondrial Redox Regulation

How is the oxidation-reduction state of mitochondria maintained during anoxia? Although this is a fundamental question, there is remarkably little information with which to answer it. It appears to be an overlooked aspect of anaerobic metabolism and in need of much further study. Tentatively, however, three possibilities can be suggested. First, 3-hydroxybutyrate dehydrogenase function could contribute to NADH regulation, with 3-hydroxybutyrate accumulating as an anaerobic end product.

Second, a reversal of β-oxidation or chain elongation, probably feasible only if electron transfer and phosphorylation are uncoupled, could contribute to redox stabilization. In this event, for each mole of acetyl coenzyme A, incorporated into fatty acyl CoA, two reducing equivalents are required and the process could contribute to maintaining redox balance. Lipids formed under these conditions may be formally viewed as anaerobic end products. Although the quantitative significance of this process if not known, there seems to be strong evidence that the process occurs in hypoxic or anoxic stress.

The third and best-documented mechanism contribu-

Table 1.3 Relationship between NADH used and succinate formed by Tubifex submitochondrial preparations.

	μmoles	
Exp. No.	NADH	Succinate
1	5	4.62
2	5	4.74
3	5	4.38

Source: Schottler (1977).

ting to mitochondrial redox regulation involves reversal of the Krebs cycle with succinate serving as a kind of carbon sink. Succinate can be formed either by an energy-linked reductase step using fumarate as an electron acceptor, or it can be formed from succinyl coenzyme A, a process catalyzed by succinate thiokinase. In vivo, precursors for these processes in vertebrates are not well known, although aspartate, malate, and fumarate added exogenously will prime the fumarate reductase, while glutamate or 2-ketoglutarate will prime the succinate thiokinase route to succinate.

The formation of succinate by fumarate reductase proceeds ultimately with the oxidation of NADH, while the formation from glutamate of 2-ketoglutarate reduces NAD^+ (table 1.3). Thus the use of succinate as a true anaerobic end product potentially contributes to mitochondrial redox balance. As it turns out, succinate accumulation proceeds with the production of molar equivalents of ATP. Thus this pro-

Table 1.4 Phosphorylation coupled to the formation of succinate.

Added malate	DNP	Succinate formed	ATP formed	ATP/succinate
5	0	3.3	1.2	0.35
5	0	3.5	1.4	0.41
5	0	3.7	1.8	0.50
5	1.5	3.9	0.3	0.09
5	1.5	3.9	0.3	0.09
5	1.5	3.7	0.2	0.07

Source: Schottler (1977), using Tubifex preparations under anoxic conditions.

cess, which can be inhibited by uncouplers, also influences the energy yield of anaerobic metabolism (table 1.4).

ATP Sources during Anoxia

In some sense, the raison d'être of anaerobic metabolism is ATP production. Since ATP is the cell's energy currency, it is important to emphasize the nature of the coupling between metabolism and ATP production. During anoxia the fermentation of glucose proceeds with the concomitant formation of 2 moles of lactate per mole of glucose, and 2 moles of ATP form from adenosine diphosphate (ADP) and orthophosphate (P_i). For oxidative metabolism, glucose oxidation to CO_2 and water is coupled to the synthesis of 36 moles of ATP per mole of glucose. But why 2? Why 36? These are difficult and perhaps unanswerable questions. However, we can say one thing with certainty: Although the fermentation of 1 mole of glucose to 2 moles of lactate is an obligate reaction stoichiometry, fixed by chemical necessity, there are no chemical or physical factors dictating that glucose fermentation to lactate must proceed with the formation of 2 moles of ATP per mole of glucose. That is, adenylate coupling to glycolysis or indeed to any metabolic sequences is an evolved coupling stoichiometry fixed by biological fitness and biological adaptation.

But what functional factors determine adenylate coupling stoichiometry? Again the question may be good, but the answers are not fully clarified. One possibility, suggested by Atkinson (1977), is that evolved coupling stoichiometries are a compromise between energy (ATP) yield and competitive ability. The nature of this compromise is illustrated in figure 1.2. It indicates that the overall conversion of glucose to lactate when uncoupled from ATP formation proceeds with a large free energy drop and an equilibrium constant of about 10^{39}. Coupling the production of 2 moles of ATP per mole of glucose fermented maintains a high capacity for competing for limiting substrate, but somewhat less than half as much ATP is made as is theoretically possible. Indeed, figure 1.2 indicates that even if the production of 4 moles of ATP were coupled to the fermentation of 1 mole of glucose to lactate, the overall equilibrium

$$\text{glucose} \longrightarrow 2 \text{ lactate}$$

$$\text{nADP} + \text{nP}_i \qquad \text{nATP}$$

n value	K	Energy yield	Competitive capacity
0	10^{39}	0	very high
2	10^{23}	low	high
3	10^{15}	medium	medium
4	10^7	high	low
5	near 1	maximal	minimal
6	10^{-8}	negative (reverse process)	low

Figure 1.2 Adenylate coupling to glycolysis: a compromise between energy yield and competitive capacity. (See Atkinson 1977 for further discussion of evolved coupling stoichiometry.)

would still be somewhat favorable. However, competitive ability for limited substrate would be drastically reduced. It is evident that whatever the factors were that actually determined the stoichiometry of adenylate coupling to glycolysis, neither ATP yield nor competitive capacity was maximized. Rather, both were compromised. And in the compromise between energetics and kinetic efficiency (undoubtedly set in primeval time) the overall biological function was optimized.

Despite these considerations, a question remains. Why is the equilibrium constant for energy metabolism so high? At least two reasons can be readily offered, one biochemical, the other biological. The biochemical reason stems from the fact that equilibrium constants of regulatory enzymes are large. Thus such enzymes can be held far from equilibrium. The product of several such large K values (equilibrium values) leads to very large overall equilibrium constants. The biological reason for maintaining high K values clearly is that this favors function at low substrate concentrations thus favoring unidirectionality of metabolic sequences, and unidirectionality greatly simplifies metabolic control.

Coupling Intermediates

The above three primary requirements for anaerobic metabolism (stored substrate, redox balance, ATP yield) would appear adequate for temporary periods of anoxia. But they are wholly inadequate if anoxia is to be sustained for periods long enough to require significant biosynthetic processes. This idea can be well expressed by a functional block diagram (fig. 1.3) showing that metabolism can be divided into functional units interconnected by a remarkably small number of mechanisms. The functional blocks for catabolism and for cellular work are coupled in three ways:

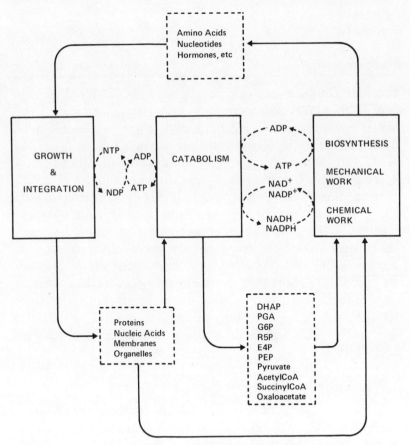

Figure 1.3 A functional block diagram of metabolism in an eucaryotic cell. Although metabolic details are omitted, the diagram emphasizes functional interactions and the small number of connections between the blocks.

Table 1.5 Intermediates for coupling different functional units of metabolism.

Metabolites needed	Supplied by glycolysis
Triose-P	yes
Tetrose-P	no
Pentose-P	no
Hexose-P	yes
Phosphoenolpyruvate	yes
Pyruvate	yes
Acetyl CoA	no
2-ketoglutarate	no
Succinyl CoA	no
Oxaloacetate	no
Minimum total: 10	Maximum suppply: 4

through the adenylates, through the $NAD^+/NADH$ and $NADP^+/NADPH$ redox couples, and through a small number of key intermediates generated by the catabolic block. A minimal estimate of the number of such intermediates is about ten, not counting isomeric or metabolically fully interconvertible compounds. When a list of such key coupling intermediates is closely examined, it is evident that anaerobic glycolysis maximally can supply only four of these (table 1.5). Thus during extended anoxia or severe hypoxia, there may be a need for auxiliary metabolic processes to augment the small pool of coupling intermediates formed by glycolysis. Moreover, from these considerations we can conclude that anaerobic glycolysis in its simplest form would not be a good strategy for organisms such as parasitic helminths, which must be capable of sustaining anoxia indefinitely.

Potentially Adjustable Factors

Most students of life sciences are taught a very conservative view of anaerobic metabolism that emphasizes uniformity and stability of the primary mechanism. But such a view can oversimplify many aspects of the process; and an even more serious deficiency is that it yields no insight into what can and what cannot change in anaerobic metabolism when O_2 availability changes. At least on the latter question, a comparative study of how various organisms deal with low O_2 problems can greatly broaden our perspective. In this broader perspective the metabolic provisions that are requisite for animal life without O_2 are not seen as unchangeable, but rather as the very raw material of adaptation, all potentially modifiable and hence potentially sensitive to selective pressure.

A minimal list of such potentially adjustable provisions for anaerobic metabolism would include:

1. a fermentable storage form of energy,
2. mechanisms for maintaining redox balance in cytosol and mitochondria,
3. mechanisms for generating different kinds and amounts of end products,
4. mechanisms for adjusting adenylate coupling stoichiometry, and
5. mechanisms for controlling the numbers and kinds of coupling metabolites.

In the first chapter, we have seen the known range of possible changes in these parameters. In the chapters that follow, individual groups of animals will illustrate ways in which the primary needs of anaerobic metabolism can be met, ways in which anaerobic metabolism (and thus hypoxia tolerance) can be improved, and functions played by auxiliary anaerobic mechanisms.

Suggested Readings

ATKINSON, D. E. 1977. *Cellular Energy Metabolism and Its Regulation.* New York: Academic Press.

CASCARANO, J., ADES, I. Z., and O'CONNOR, J. D. 1976. Hypoxia: a succinate-fumarate electron shuttle between peripheral cells and lung. *J. Exp. Zool.* 198:149–154.

DAVIDOVITZ, J., PHILIPS, G. H., PACHTER, B. R., and BREININ, G. M. 1975. Cisternal distention in membrane-glycogen complexes of rabbit extraocular muscle. *J. Ultrastruct. Res.* 51:307–313.

FIELDS, J. H. A., BALDWIN, J., and HOCHACHKA, P. W. 1975. On the role of octopine dehydrogenase in cephalopod mantle muscle metabolism. *Can. J. Zool.* 54:871–878.

FIELDS, J. H. A., HOCHACHKA, P. W., ENG, A. K., RAMSDEN, W. D., and WEINSTEIN, B. 1979. Alanopine and strombine are novel imino acids produced by a dehydrogenase found in the adductor muscle of the oyster, *Crassostrea gigas*. *Arch. Biochem. Biophys.*, in press.

HOCHACHKA, P. W., 1976. Design of metabolic and enzymatic machinery to fit lifestyle and environment. *Biochem. Soc. Symp.* 41:3–31.

HULBERT, W. C., GUPPY, M., MURPHY, B., and HOCHACHKA, P. W. 1979. Metabolic sources of heat and power in tuna. I. Muscle Fine Structure. *J. Exp. Biol.*, 82:289–301.

HOCHACHKA, P. W., and HULBERT, W. C. 1978. Glycogen 'seas', glycogen bodies, and glycogen granules in heart and skeletal muscles of two air-breathing, burrowing fishes. *Can. J. Zool.* 56:774–786.

KEMP, J. D. 1977. A new amino acid derivative present in crown gall tumor tissue. *Biochem. Biophys. Res. Commun.* 74:862–868.

SCHOTTLER, U. 1977. The energy-yielding oxidation of NADH by fumarate in anaerobic mitochondria of *Tubifex* sp. *Comp. Biochem. Physiol.* (B) 58:151–156.

STOREY, K. B., and STOREY, J. M. 1979. Octopine metabolism in the Cuttlefish, *Sepia officinalis:* octopine production by muscle and its role as an aerobic substrate for non-muscular tissues. *J. Comp. Physiol.*, (B) 131:311–320.

Chapter Two

Helminths and the Usefulness of Carbon Dioxide

When the availability of oxygen becomes severely limited to the cell, the tissue, or the organism, the common response is to activate anaerobic metabolism to compensate for the temporary lack of oxygen. This is the metabolic response of a person's skeletal muscles when he sprints after a bus, and it is limited here, as elsewhere, by the need to repay an oxygen debt. Not surprisingly, the best of animal anaerobes, the parasitic helminths, try to avoid this limitation and therefore rarely utilize anaerobic glycolysis in its simplest form. Instead, these champion anaerobes have evolved metabolic machinery that has allowed them to survive in often hostile environments such as the mammalian gastrointestinal tract and exploit its rich nutrient sources. Interestingly, gastrointestinal parasites have developed the most effective anaerobic metabolism known despite the periodic occurrence of oxygen in the intestinal lumen, implying a complex set of selective forces acting in this environment.

That complexity is strikingly illustrated by the multiple anaerobic end products that can be produced. The commonest end products are probably succinate, propionate, acetate, three longer-chain volatile fatty acids, and carbon dioxide. Lactate may or may not accumulate, suggesting that lactate dehydrogenase (LDH) may or may not be present.

But why multiple end products? By what reaction mechanisms are these formed? What are their functions?

These kinds of questions were already being pondered over thirty years ago, for the complexity of anaerobic end products in parasitic helminths was well recognized by Von Brand in his classic treatise of 1946. Although of major importance, recognition of the accumulation of novel anaerobic end products did not of itself seem to supply sufficient clues to adequately explain how metabolism worked in these organisms. Something was still missing. That something was carbon dioxide fixation. It turned out that in parasitic helminths, CO_2 plays a special role, in that fermentative metabolism often proceeds with the concomitant fixation of molecular carbon dioxide. That is, in these organisms, CO_2 is both fixed (as a potential substrate source) and released (as a potential anaerobic end product) during fermentation.

Where and How CO_2 Is Fixed

The observations of CO_2 fixation in helminth worms go back a long way and stem from studies showing (1) the dependence of growth and metabolism on CO_2, (2) the disappearance of CO_2 from anaerobic bicarbonate media, and (3) the dependence of glycogen usage on CO_2. In 1954 Fairbairn showed that CO_2 was fixed under anaerobic conditions and appeared in propionate and probably succinate. These pioneering efforts served as the point of departure for metabolic studies of helminths over the last fifteen to twenty years. Carbon dioxide fixation is now known to be an integral component of anaerobic metabolism in trematodes (example, *Fasciola*), in nematodes (example, *Ascaris*), and in cestodes (example, *Hymenolepis*), as well as in many other invertebrate groups.

Where, in overall metabolism, the CO_2-fixing step occurs can be well illustrated by considering the pathway of glucose fermentation in *Hymenolepis* (fig. 2.1). To the level of phosphoenolpyruvate (PEP), the pathway appears to be qualitatively similar or identical to that in higher vertebrates. The enzyme levels in the glycolytic path are not at all unusual and indicate that either glycogen or glucose can serve as substrates for the pathway. But at the phosphoenolpyruvate level, a novel situation is found. Pyruvate kinase

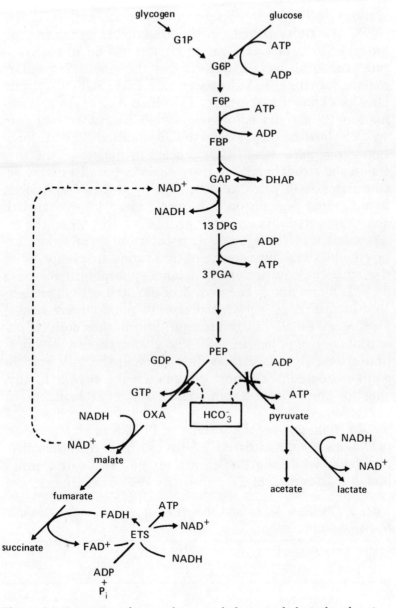

Figure 2.1 Summary of anaerobic metabolism in helminths showing probable routes to succinate, acetate, and lactate.

activity is low, but phosphoenolpyruvate carboxykinase (PEPCK) activity is high. In the presence of a divalent cation, PEPCK catalyzes a carboxylation of PEP to oxaloacetate; the oxaloacetate formed can then be converted to malate, fumarate, and ultimately succinate, which accumulates as a known anaerobic end product. As a rough approximation, PK activity is low while PEPCK activity is high in species showing most effective CO_2 fixation. PEPCK function explains (1) why $^{14}CO_2$ appears in malate as the first stable and strongly labeled intermediate, but ultimately accumulates mainly in succinate; (2) why glucose or glycogen fermentation depends on CO_2; and, indeed, (3) why growth and CO_2 availability are so linked. In *Ascaris,* when the intermediates of the glucose fermentation pathway are assayed and metabolite concentration ratios are compared to the apparent equilibrium constants, phosphofructokinase (PFK) is found not to be in equilibrium, and is thus predictably identified as a potential control point. Interestingly, PEPCK and PK also are not in equilibrium; thus both are potential control points in anaerobic glycolysis and the question arises: How is carbon flow through the PEP branchpoint regulated? The most comprehensive answer to this question comes from recent studies of the PEP branchpoint in *Hymenolepis.*

In *Hymenolepis* the activity of PEPCK at physiological pH values is approximately 2.5 times higher than the activity of PK, while the PK affinity for PEP is about 4 times higher than is the PEPCK affinity for PEP. Thus, the ratio of

Table 2.1 Pyruvate kinase and phosphoenolpyruvate carboxykinase in *Hymenolepsis.* [a]

Catalytic function	PK	PEPCK
$K_{m(PEP)}$	0.1	0.4
V_{max}	0.1	0.25
$\dfrac{V_{max}}{K_m}$	1.0	0.6

Source: Podesta et al. (1976).

[a] Assay at pH 7.4, 37°C. V_{max} in μmoles/min/mg protein. K_m in mM.

V_{max}/K_m for the two enzymes is similar (table 2.1), indicating a potential split in carbon flow through this branchpoint at physiological pH and physiological substrate concentrations.* Yet in the postprandial intestine, *Hymenolepis* accumulates succinate as the dominant end product, while between meals lactate is the predominant end product. This implies that the two enzymes, PK and PEPCK, are closely regulated and the question is how.

The basic problem of turning off PK, thereby allowing PEPCK activity and thus carbon flow to succinate, seems solved by the CO_2/bicarbonate system. In vivo, in the postprandial intestine, CO_2 builds up to 600 mm Hg. Carbon dioxide diffuses into the worm, shifting the reaction,

$$CO_2 + H_2O \rightleftharpoons H^+ + HCO_3^-$$

to the right. Calcareous corpuscles are stored in the worm and these may be simultaneously mobilized to buffer the hydrogen ion, thus leading to a further buildup of bicarbonate. This buildup of bicarbonate serves to activate PEPCK strictly as a substrate saturation phenomenon and serves to simultaneously inhibit pyruvate kinase by a drastic reduction in PK affinity for PEP. The sum of these conditions favors the carboxylation of PEP to form oxaloacetate which ultimately accumulates as succinate, the dominant end product accumulated at this time. Between meals, when CO_2 in the lumen is reduced, the worm is known to retreat to lower intestinal regions, where the reverse of the above regulatory interactions is thought to occur. Thus the reduction in CO_2 deinhibits PK, while simultaneously it reduces PEPCK activity. Under these conditions, carbon flow in the direction of pyruvate, and ultimately lactate, is favored. As lactate builds up, it tends to augment its own further production by an inhibitory effect upon PEPCK. This arrangement adequately accounts for succinate formation, lactate formation, and CO_2 fixation, but it does not explain the origin of volatile fatty acids often produced by helminth worms.

* V_{max} is the maximum velocity in μmoles/min/mg enzyme, K_m is the Michaelis-Menten constant, the concentration of substrate at which the reaction velocity $(V) = \frac{1}{2}V_{max}$.

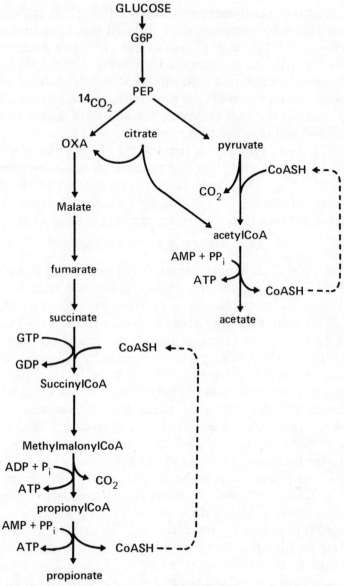

Figure 2.2 Metabolic map showing the probable origins of acetate and propionate in helminths.

Origin of Propionate and Acetate

Results from five key experiments bear on this question. They are as follows:

1. ^{14}C glucose appears in propionate and acetate;
2. ^{14}C pyruvate appears in acetate but not propionate;
3. $^{14}CO_2$ appears in propionate but not in acetate;
4. 2,3-^{14}C-labeled succinate appears in propionate but not acetate;
5. 1,5-^{14}C-labeled citrate appears in acetate.

These and similar results are consistent with the usual pathway of glucose metabolism to the level of PEP (fig. 2.1). Acetate is thought to be formed from pyruvate via acetyl CoA, while succinate is thought convertible to propionate via their respective CoA derivatives (fig. 2.2). Although further work is required in this area, the current explanations adequately account for the formation of propionate and acetate from glucose carbon. In *Fasciola*, at the same time, however, longer-chain volatile fatty acids are also released, and these receive no glucose carbon. Thus longer-chain volatile fatty acids must have an alternate origin. Branch-chain amino acids are considered the most likely candidates.

Origin of Volatile Fatty Acids

The key experimental observations here are as follows: (1) ^{14}C leucine is incorporated mainly into isovalerate; ^{14}C isoleucine is incorporated mainly into methylbutyrate; ^{14}C valine is incorporated mainly into isobutyrate. (2) All the above processes are stimulated by exogenous glucose (table 2.2). (3) The production of each volatile fatty acid varies with the concentration of the appropriate branch-chain amino acid; thus isovalerate, for example, varies directly with the leucine concentration, at the same time, propion-

Table 2.2 Glucose effect on leucine fermentation to isovalerate in *Fasciola*.

Conditions	Isovalerate formed	Specific activity
No glucose	13	39
+200 μM glucose	21	41

Source: Lahoud et al. (1971).

Table 2.3 Effect of leucine on isovalerate formed and on the concentration ratio of propionate/acetate in *Fasciola*.

Leucine concentration	Isovalerate formed	Propionate/Acetate
10 μM	1.7	2.1
400 μM	18.2	7.3

Source: Lahoud et al. (1971).

ate production increases (table 2.3). (4) Leucine enhances ^{14}C glucose incorporation into propionate. (5) All these interactions are abolished by high oxygen. All of these results are consistent with the concept of branch-chain amino acid fermentation to volatile fatty acids in these helminths. The probable pathways (fig. 2.3) involve three linked phases: (1) transamination of branch-chain amino acids to the appropriate keto acids catalyzed by specific aminotransferases; (2)

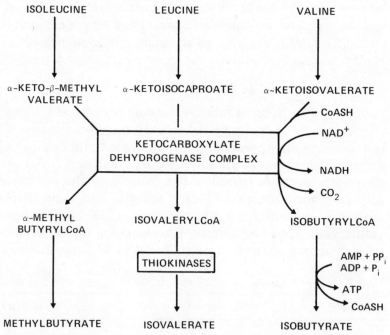

Figure 2.3 Proposed scheme for the fermentation of branched chain amino acids in helminths.

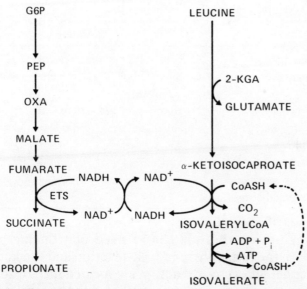

Figure 2.4 Possible connection between glucose and branched-chain amino acid metabolism.

conversion of the keto acids to coenzyme A derivatives, reactions probably catalyzed by a single keto carboxylate dehydrogenase complex with a generalized keto acid specificity; and (3) conversion of the CoA derivative to the appropriate fatty acid, with the regeneration of CoA and the coupled formation of ATP. This model is consistent with all currently available data, and is particularly satisfying in that it identifies at least one origin of true metabolic CO_2 generated in phase 2 of branch-chain amino acid fermentation (that is, by the keto carboxylate dehydrogenase reactions). However, it does not explain the above-mentioned connection between glucose fermentation to propionate and branch-chain amino acid fermentation to volatile fatty acids.

One possible explanation for the apparent connection between glucose fermentation to propionate and leucine fermentation to isovalerate is shown in figure 2.4. The linear path of glucose to succinate and hence propionate is by itself out of redox balance, as is the linear path of leucine to isovalerate fermentation. One of the pathways is in need of

Table 2.4 Hypoxia adaptations in helminths.

Requirement	Solution
Storage energy	Carbohydrate and amino acids
Redox regulation	Coupled carbohydrate and amino acid catabolism
CoASH regulation	1:1 function of ketocarboxylate dehydrogenase + thiokinases
ATP yield	
Glucose → propionate	6
Leucine → isovalerate	1
Coupling metabolites [a]	8

[a] Ten ideally needed.

oxidizing power, the other is in need of reducing power. Since both the fumarate reductase reaction and the keto carboxylate dehydrogenase reaction are located intramitochondrially, it is tempting to suggest that a redox couple forms between them. That certainly would nicely account for the observed experimental correlation between glucose conversion to propionate and leucine conversion to isovalerate. As we shall see, this is but one example of coupling between carbohydrate and amino acid fermentation.

What are the functional consequences of this metabolic organization? What are the metabolic advantages? We can perhaps best assess this by considering how the basic requirements for sustained anoxia in the helminths are achieved (table 2.4).

Hypoxia Adaptations in the Helminths

As emphasized in chapter 1, a key requirement for sustained anoxia is a storage form of carbon and energy. In classical glycolysis, this is glycogen, while in the helminths, two substrate sources appear to be utilized: carbohydrate and amino acids. Another major requirement for sustained anoxia is redox regulation, both in the cytosol and the mitochondria. Parasitic helminths utilized a novel solution to this problem in that both carbohydrate and amino acid fermentations are required to balance redox. Neither process appears to be independently functional. A third requirement for sustained anoxia in parasitic helminths relates to the regulation of CoA reserves, since these must neither be depleted nor accumulated during sustained anaerobic metabolism. Although in theory this is a problem that will be faced in any cell, vertebrate or invertebrate, sustaining anoxia, it does not seem to be as much involved in higher organisms. It becomes a critical problem in the parasitic helminths because of the fermentation of branch-chain amino acids. Depletion or accumulation of CoA may be prevented through a 1:1 functional coupling of keto carboxylate dehydrogenase and thiokinases.

A fourth requirement for sustained anoxia in invertebrates, underpinning all others, is of course the generation of ATP. As is evident from fig. 2.1–2.3, the fermentation of glucose through to propionate is capable of yielding 6 moles of ATP per mole of glucose. Simultaneously, if this is coupled to leucine fermentation to isovalerate, a mole of ATP is generated per mole of leucine. The total yield of ATP is thus minimally 7 moles of ATP per mole of glucose plus branch-chain amino acid fermented. By comparison, the classical glycolytic system produces 2 moles of ATP per mole of glucose. Hence energetic efficiency is increased by a substantial factor.

Finally, for sustained anoxia the generation of coupling metabolites utilized for biosynthetic and other essential metabolic functions is required. By comparison with anaerobic glycolysis, which is able to generate 4 such essential coupling metabolites, the anaerobic metabolism of parasitic helminths is minimally capable of generating at least 8 of the 10 ideally needed components. Perhaps more so than any other factor, it is this increased capacity to generate key coupling intermediates that renders helminth anaerobic metabolism so surprisingly versatile and flexible. It is all the more exciting, therefore, to realize that other free-living organisms have invented a similar metabolic organization.

Suggested Readings

BARRETT, J., and BEIS, I. 1973. Studies on glycolysis in the muscle tissue of *Ascaris lumbricoides* (Nematoda). *Comp. Biochem. Physiol.* (B) 44:751–761.

BRYANT, C. 1975. Carbon dioxide utilization and the regulation of respiratory metabolic pathways in parasitic helminths. In Dawes, B., ed., *Advances in Parasitology*, vol. 13. New York: Academic Press. Pp. 36–69.

LAHOUD, H., PRICHARD, R. K., MCMANUS, W. R., and SCHOFIELD, P. J. 1971. The dissimilation of leucine, isoleucine, and valine to volatile fatty acids by adult *Fasciola hepatica. Intl. J. Parasitol.* 1:223–233.

PODESTA, R. B., MUSTAFA, T., MOON, T. W., HULBERT, W. C., and MET-TRICK, D. F. 1976. Anaerobes in an aerobic environment: role of CO_2 in energy metabolism of *Hymenolepis diminuta.* In Van den Bossche, ed., *Biochemistry of Parasites and Host-Parasite Relationships.* New York: North-Holland. Pp. 81–88.

SCHMIDT, G. D., and ROBERTS, L. S. 1977. *Foundations of Parasitology.* St. Louis: Mosby.

VON BRAND, T. 1946. *Anaerobiosis in Invertebrates.* Biodynamica Monographs, 4. Normandy, Missouri: Biodynamica.

———. 1966. *Biochemistry of Parasites.* New York: Academic Press.

Chapter Three

Coupled Glucose and Amino Acid Catabolism in Bivalve Molluscs

It is a little-appreciated fact that bivalve molluscs have an unusually high tolerance to anoxia. The oyster (*Crassostrea*), mussel (*Mytilus*), and mud clam (*Rangia*) are three of the best-studied species. The time course of anoxia response in the oyster seems to be typical. Within a short time of shell closure, the oxygen tension in the pallial fluid drops to a level of about 40 mm Hg for the duration of the anoxic period (over 3 weeks at 25°C). The pH of the pallial fluid also changes in a characteristic way. Within the first hour, the pH drops from 7.6 to about 6.7, which is the pK for bicarbonate buffer. For the subsequent 23 days at 25°C, no further sharp pH drop occurs, but there is a very slight shift downward. At about 23 to 24 days, the pH drops again quite precipitously to 5.4, a value near the pK for succinate but still substantially above the pK for propionate (pH 4.8). Under these conditions, the oyster survives for only a few more days.

As in the helminths, multiple anaerobic end products are formed under these conditions. The proven end products include succinate, alanine, propionate, and alanopine. Succinate and alanine are probably formed during early phases of anoxia, while propionate takes on a greater and greater role in later stages of anoxia, particularly in *Mytilus*. It is probable that is also true for alanopine. Less common or suspected anaerobic end products include lactate, acetate, and other volatile fatty acids. Their relevance to molluscan metabolism, however, will only be established with further studies.

Again as in the helminths, there is a dual role for carbon dioxide. During anaerobic metabolism, the evidence indicates that CO_2 is both fixed and produced. During early phases of anoxia, the ratio of CO_2 fixation to CO_2 production perhaps is similar, and for the first seven or eight days of anoxia the partial pressure of CO_2 (PCO_2) as a result does not change. However, this conclusion could be modified by the fact that CO_2 levels could also be influenced by mobilization of calcium carbonate from the animal's shell. After the first week of anoxia, PCO_2 begins to rise in the pallial fluid. This rise may be caused by acidic end products being formed, but since the pH change at this time is modest, the PCO_2 rise may be due to metabolic factors; that is, anaerobic CO_2 production may exceed anaerobic CO_2 fixation in these extreme stages of anoxia.

Fermentative Pathways of Metabolism

The fermentative pathways in bivalve molluscs that appear to account for the kinds of end products formed and their patterns of formation are shown in fig. 3.1. Many features of these pathways appear to be similar to those in helminths. Thus the fermentation of glucose to the level of phosphoenolpyruvate follows the standard glycolytic path. A distinct metabolic branchpoint may arise at the level of PEP, which may be converted either to oxaloacetate via PEP carboxykinase or to pyruvate, a reaction catalyzed by pyruvate kinase. The extent of competition at this metabolic branchpoint of course depends on the relative activities of PEPCK and PK. When both enzymes occur at significant levels in the same tissue, their activities appear to be regulated by fructose biphosphate (FBP), alanine, pH, and the energy status of the cell. As a result, significant simultaneous function of the two enzymes would occur only during transient metabolic states, on moving from conditions favoring one to those favoring the other of the PEP branchpoint reactions. Whatever the precise kinetic control mechanisms are at the PEP branchpoint, it is evident that there are at least two linear paths of glucose fermentation in the bivalves, one leading to alanine accumulation and the other leading

to succinate accumulation, with the succinate being ultimately convertible to propionate if anoxia is extreme and
extended. In fact, under these conditions propionic acid
may account for substantially over 50% of the carbon accumulated in anaerobic end products.

These kinds of metabolic schemes, which are really
summaries of our current knowledge of anaerobic metabolism in these organisms, are largely based on whole organism
work. This has been convenient, because bivalves used in
experimentation are usually small and their circulatory systems are difficult to monitor. But such approaches can lead
to artifacts or rather unexpected results. That is why recent
studies of the isolated anoxic oyster heart have been useful
in clarifying the complicated anaerobic metabolism found
in these animals.

The isolated oyster heart was chosen for several metabolic features. Because it consists of a single cell type with
basically a single physiological function, it is a simple system. Although the heart can tolerate anoxia, heart cells contain abundant mitochondria. As a direct result, this organ
clearly must undergo a true aerobic/anaerobic transition
when the whole organism enters anoxia. Despite these char-

Table 3.1 Concentrations of some metabolites in the oyster ventricle
before and after anaerobiosis.

| | Metabolite concentration (μmol/g wet wt.) | | | |
| | In vitro | | In vivo | |
Metabolite	Aerobic	Anaerobic	Aerobic	Anaerobic
Oxaloacetate	<0.08	<0.08	<0.08	<0.08
Citrate	1.84	0.96	1.04	1.52
α-ketoglutarate	<0.08	<0.08	<0.08	<0.08
Pyruvate	0.16	0.08	0.08	0.12
L-lactate	0.24	0.32	0.0	0.28
Malate	0.24	0.48	0.08	0.24
Succinate	0.56	2.72	0.16	2.52
Aspartate	10.16	1.12	8.40	4.32
Glutamate	8.16	5.52	8.00	10.24
Alanine	5.36	9.52	7.04	14.52

Source: Collicutt and Hochachka (1977).

acteristics, the oyster heart has little or no LDH activity. Finally, and equally significantly, PEP carboxykinase activity in the oyster heart is so low it is hardly measureable. The deletion of both LDH and PEPCK from the oyster heart indicates that neither lactate nor succinate could form as predominant anaerobic end products from glucose. What end products, then, are produced?

Metabolite changes occurring when the isolated oyster heart is incubated under anoxia conditions are shown in table 3.1. They indicate that succinate and alanine are the predominate end products accumulating. Similar results occur when the whole organism is put under anoxia, the heart is freeze-clamped, and measurements are made of anaerobic end products formed. Alanine under both conditions is presumably formed from pyruvate by alanine aminotransferase, but in the absence of PEPCK how is succinate formed? The answer: not from glucose.

Origin of Succinate in Anoxic Oyster Heart

One of the surprising observations of metabolite profiles in the oyster heart is the occurrence of high (8 to 12 μmol/gm) aspartate concentrations. During anoxia, aspartate levels drop both in the isolated anoxic heart and in the heart of the anoxic oyster. The falling levels of aspartate in experiments with whole organisms are not as drastic as in the isolated oyster heart preparations, indicating the possibility of regneration and supply of aspartate from other organs. Be that as it may, the data imply that aspartate may be the ultimate precursor (storage substrate source) for the anaerobic production of succinate.

Experiments with ^{14}C aspartate establish that linear path with certainty. Of the ^{14}C aspartate carbon taken up, about half accumulates in succinate under anoxic conditions, while another 20% occurs in malate presumably enroute to succinate (table 3.2).

The forward flow of carbon through a segment of the Krebs cycle, from 2-ketoglutarate (2-KGA) or metabolites convertible to 2-KGA, supplies another possible route to succinate (fig. 3.1). This indeed was originally postulated as a significant source of succinate. In the anoxic oyster heart about 50% of the ^{14}C glutamate carbon taken up is incor-

Table 3.2 Metabolism of ^{14}C-labeled glucose and amino acids by the isolated anoxic ventricle of the oyster.

Process	Glucose ^{14}C(U)	Aspartate ^{14}C(U)	Glutamate ^{14}C(U)	Alanine 1-^{14}C
Uptake of ^{14}C-labeled compound by tissue (% of added tracer)	16.4	9.3	5.3	6.0
% intracellular ^{14}C as ^{14}C-precursor	0.0	30.3	85.4	77.3
% of metabolized precursor ^{14}C in:				
Glutamate	—	2.1	n.a.[a]	—
Aspartate	—	n.a.	10.1	—
Other amino acids	—	—	28.3	—
Unknown compound[b]	30.4	11.3	trace	67.1
Alanine	55.2	16.7	4.8	n.a.
Phosphorylated glycolytic intermediates	5.3	—	—	—
Pyruvate	1.0	2.2	—	32.9
Malate	5.5	19.7	10.7	—
Succinate	2.7	48.1	46.2	—

Source: Collicutt and Hochachka (1977), with modification.
[a] Not applicable.
[b] Identified by J. H. A. Fields as alanopine.

porated into succinate. Thus the overall impression can be gained of succinate as a carbon sink for the flow of materials down two branches of the Krebs cycle. In one branch (oxaloacetate to succinate) the flow of carbon is in the reverse direction, while in the other branch (2-KGA to succinate) it is in the forward direction compared to Krebs cycle function under aerobic conditions. However, the absolute amount of glutamate taken up for this function is so low that it is assumed to represent a trivial contribution to overall anaerobic metabolism. That leaves us with aspartate fermentation to succinate as a linear metabolic pathway clearly out of redox balance in the cytosol. It is probably held in redox balance by the simultaneous fermentation of glucose.

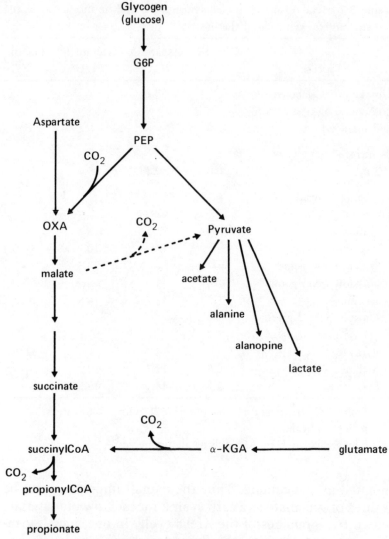

Figure 3.1 Overall scheme of anaerobic metabolism in bivalves.

End Products Formed from Glucose

When the anoxic oyster heart is incubated in a medium containing ^{14}C glucose, less than 5% of the ^{14}C glucose carbon appears in succinate, consistent with the near absence of PEPCK. In contrast, alanine serves as the primary carbon sink for glucose, about 55% of the ^{14}C glucose taken up being incorporated into alanine. This linear path of glucose fermentation to alanine is also clearly out of redox balance. For its continued function NAD^+ could be regenerated by the malate dehydrogenase (MDH) reaction, primed by aspartate conversion to oxaloacetate, and to some extent this must occur. But if this were the only redox mechanism, it would imply that dropping aspartate concentrations should be stoichiometrically coupled with the rise in alanine concentrations in a $1:1$ ratio. In the heart of the anoxic oyster this stoichiometry does not occur, implying that something is still missing, some other mechanism must occur. That other mechanism involves the formation of what was originally an unknown compound and accounted for some 30% of the ^{14}C carbon of glucose utilized during anoxia.

Alanopine as an Anaerobic End Product

Several features of the unknown end product formed from ^{14}C glucose gave the first clues as to its identity. (1) The unknown was a single compound, and the same compound could be formed from glucose, alanine, aspartate, or pyruvate as precursors. (2) The unknown did not co-chromatogram with any of the natural amino acids or with taurine, citrulline, methionine, or octopine. (3) Acid hydrolysis showed that it was not a peptide. The unknown compound bound to ion exchange columns in a manner similar to amino acids and other nitrogen-containing compounds such as octopine, indicating the occurrence of a net positively-charged nitrogen group. (5) The amount of unknown produced from ^{14}C glucose varied with the amount of ^{14}C alanine produced. (6) With ^{14}C alanine as a precursor, only the unknown and pyruvate were formed. From these clues, Fields postulated that the unknown compound was N-carboxyethylalanine (alanopine) formed by the reductive condensation of pyruvate and alanine:

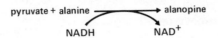

Formal proof for this hypothesis required the isolation of the appropriate enzyme, the stoichiometric utilization of substrates and coenzymes, the demonstration of the back reaction, the identity of the product formed in vivo and in the isolated enzyme reaction, the identity of the product formed in the enzyme reaction and that formed de novo by organic synthesis, and, finally, structural identity of the compound synthesized by the enzyme reaction and by organic synthesis. All of this evidence has now been marshalled and need not be reviewed here. Because this kind of reaction is not widely encountered by mammalian biochemists, however, let us briefly discuss some of its features.

Alanopine Dehydrogenase, a Small Enzyme

As part of the purification of alanopine dehydrogenase (ADH), advantage it taken of its evident separation from the bulk of oyster heart proteins by G200 Sephadex column chromatography. From calibrated columns of G200 its molecular weight is about 45,000; in size alanopire dehydrogenase is therefore similar to octopine dehydrogenase (ODH) in cephalopods. Unlike LDH, which is a tetramer of 4 subunits and a total molecular weight of about 140,000, ADH is probably a single subunit enzyme.

Substrate Specificity

Oyster heart ADH has an almost absolute specificity for pyruvate as a keto acid substrate. Of several likely alternate substrates, only 2-oxybutyrate serves as a substrate, being about 40% as effective as pyruvate.

The specificity toward alanine is not as extreme, glycine, cystine, serine, and β-alanine all being capable of serving as substrates for the reaction. Of these, however, alanine and glycine seem the most effective (tables 3.3, 3.4).

Table 3.3 Keto acid specificity of alanopine dehydrogenase.

Keto acid (3 mM)	Relative activity
Pyruvate	100.0
2-oxobutyrate	38.4
Glyoxylate	4.7
Hydroxypyruvate	4.0
Oxaloacetate	1.3
2-oxoglutarate	1.3

Source: Data from J. H. A. Fields (personal communication).

Table 3.4 Amino acid specificity of alanopine dehydrogenase.

Amino acid (200 mM)	Relative activity
Alanine	100.0
Glycine	98.0
Cysteine	50.5
Serine	50.0
β-alanine	16.0
Taurine	0
Glutamate	0

Source: Data from J. H. A. Fields (personal communication).

Kinetic Properties

In many ways ADH is analogous to LDH. For the forward reaction, the enzyme shows approximately neutral pH optima; for the back reaction, it shows strongly alkaline pH optima. The enzyme shows no regulatory properties; of a large series of glycolytic and Krebs cycle intermediates and a series of amino acids, none were found to be effective inhibitors or activators of the enzyme. The adenylates all are potentially competitive with the coenzyme binding site; since they are competitive with other dehydrogenases as well, adenylate effects on ADH lack the specificity necessary for important regulatory roles in vivo. The kinetic characteristics that seem of greatest metabolic significance are the apparent affinities for the coenzyme and the two substrates. The apparent K_m value for NADH (0.01 mM) is in the physiological range, while the apparent K_m values for alanine (about 100 mM) and pyruvate (0.4–1.0 mM) are probably higher than resting physiological concentrations. It is best evident how appropriate these kinetic constants are when it is recalled that in vivo ADH competes with MDH for the same (limiting) pool of cytoplasmic NADH.

Oyster Cytoplasmic Malate Dehydrogenase

Oyster adductor muscle malate dehydrogenase occurs in higher activities than any other enzyme in energy metabolism of this tissue, at about 120 units/g wet weight of muscle tissue, that is, about twice the level of ADH. On starch and polyacrylamide gels two electrophoretic forms are distinguishable, the cytoplasmic and mitochondrial forms of this enzyme that are known from other sources. Since mito-chondrial abundance is very low in adductor tissue, it is not surprising that most of the malate dehydrogenase activity is cytoplasmic. In the oxaloacetate $\rightarrow$ malate direction, the pH optimum for the enzyme is between pH 7.2 and 7.6, while it is alkaline (about pH 9.0) for the back reaction. Under most circumstances the enzyme follows Michaelis-Menten kinetics. The K_m values for NADH and oxa-loacetate at pH 7.5 are about 0.02 mM and 0.04 mM, respectively. As pH is lowered, the enzyme affinity for oxa-loacetate rises (K_m falls to 0.01 mM). In the back reaction, the K_m for NAD^+ and malate are 0.04 and 0.1 mM, respec-tively. Out of a long list of compounds tested, only citrate, 2-ketoglutarate, and the adenylates are found to be inhibi-tory; however, these are not considered to be physiologic-ally important modulators either because of the concentra-tions required being too high (in the case of the carboxylic acids) or because of a lack of specificity (in the case of the adenylates).

Aside from substrate affinity effects, the major determi-nant of in vivo activity probably is product inhibition. Thus it is instructive that the forward reaction is not strongly in-hibited by the products NAD^+ and malate, while the reverse reaction is potently inhibited by its products, NADH and oxaloacetate. Conditions to be expected during anaerobic metabolism, therefore, appear suitable for MDH function in the forward direction. Interestingly, malate dehydrogenase displays a lower K_m for oxaloacetate than does aspartate aminotransferase, so under most circumstances the dehy-drogenase would outcompete the transferase for limiting oxaloacetate pools. This advantage is further accentuated as pH drops during anoxia, assuring the flow of aspartate car-bon through the malate dehydrogenase step and ultimately to succinate. The catalytic and regulatory properties of MDH and aspartate aminotransferase are thus seen to be consistent with the observed fermentation of aspartate dur-ing anoxia. But the question still remains of how MDH function is integrated with the function of ADH.

Functions of Alanopine and Malate Dehydrogenases

In catalyzing the terminal step in the modified anaerobic glycolysis of bivalve adductor muscles, ADH is formally analogous to LDH in vertebrate anaerobic glycolysis. The fundamental difference is that ADH is not suited for this role during early stages of glycolytic activation when pyruvate and alanine concentrations are low. Since both these substrates for ADH are accumulated from glycogen (glucose) during anoxia, for glycolysis to proceed at this time there is a patent requirement for a source of NAD^+ for the oxidative glyceraldehyde-3-phosphate dehydrogenase (GAPDH) step in the pathway. That is why malate dehydrogenase is coupled to glycolysis in this system. Unlike ADH, MDH is quite suitable for function during early stages of glycolytic activation: it displays low K_m values for oxaloacetate and NADH, it is very active, its competitive ability rises as pH drops due to CO_2 and acid end product accumulation, and its kinetic properties favor integrated function with aspartate aminotransferase.

There is a problem with this arrangement, however, and that is aspartate availability. Although MDH is kinetically suited to supply the primary means for balancing cytosolic redox, its function can become limiting if aspartate stores are exhausted. In adductor muscle, aspartate pools are in the 5 to 10 mM range, while glycogen (glucose) is at least an order of magnitude more abundant. In oyster heart (a much more aerobic tissue), aspartate pools are in the 10 mM range, which is still lower than the glycogen (glucose) level. Whenever aspartate became limiting, therefore, anaerobic glycolysis would slow down and stop if ADH as a secondary redox control mechanism were not activated. The kinetic properties of ADH nicely fit this scheme, for ADH cannot compete well for alanine until alanine concentrations rise. However, as alanine concentrations rise, ADH activity rises, and in such conditions glycolytic activation would lead to no further alanine accumulation, but a greater and greater incorporation of glucose carbon into alanopine. This kinetic behavior would predict that alanine accumulation would not continue indefinitely during anoxia but

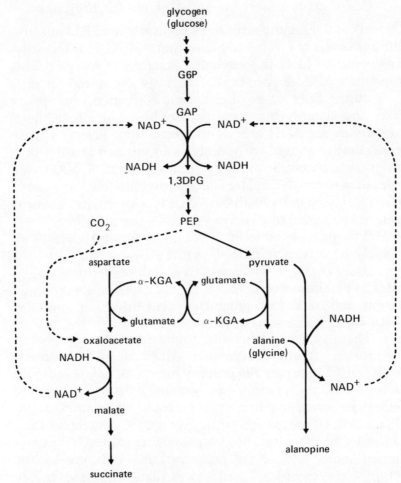

Figure 3.2 Coupling mechanisms leading to succinate, alanine, and alanopine accumulation in bivalves.

would rather approach some sort of asymptote, a prediction in satifactory agreement with available data.

Finally, the coupling of carbohydrate and amino acid fermentation in the oyster heart is assured by three mechanisms: redox coupling of GAPDH in glycolysis to MDH in the aspartate fermentation; interaction of alanine and aspartate aminotransferases through 2-KGA and glutamate tumbling between them (fig. 3.2); and direct chemical coupling through ADH, since one substrate, pyruvate, derives from carbohydrate while the cosubstrate derives either from carbohydrate or the free amino acid pool. As we shall see, this latter mechanism of direct coupling of carbohydrate and amino acid fermentation is not an isolated example, but is utilized by other organisms as well.

Hypoxia Adaptations in Bivalves

The picture emerging from the above discussion shows interesting parallels with helminths (table 3.5). As in the latter, carbohydrates and a large free amino acid pool satisfy the requirements for a storage form of energy. Redox regulation in the cytosol is achieved by at least two known mechanisms involving MDH and ADH (fig. 3.2). In the first instance, through simultaneous mobilization of glucose and aspartate, the NAD^+-linked GAPDH reaction in glycolysis is coupled to the NADH-linked MDH reaction in aspartate fermentation. The end products formed initially are alanine (from glucose) and succinate (from aspartate). As aspartate levels drop, alanine levels rise and serve to activate alanopine dehydrogenase which, by taking over cytosolic redox regulation, allows carbohydrate fermentation to continue even if aspartate pools are depleted. By these processes, a significant fraction of glucose (about 30% in anoxic oyster

Table 3.5 Hypoxia adaptations in bivalves.

Requirement	Solution
Storage energy	Carbohydrate and amino acids
Redox regulation	Glucose + aspartate coupled metabolism
	Pyruvate + alanine for alanopine dehydrogenase
	Fumarate reductase?
CoASH supplies	Succinate thiokinase + propionate thiokinase
	Keto carboxylate dehydrogenases + thiokinases
ATP yield	
Glucose → propionate	6
Aspartate → succinate	1
Other thiokinases?	
Coupling metabolites[a]	8

[a] Shunt metabolites may be missing.

hearts) appears in alanopine, which represents a true end product of anaerobic glycolysis.

Unlike the situation in the cytosol, mitochondrial redox regulation in bivalves is poorly understood. However, it is clear that succinate accumulation is due largely to fumarate reductase and this process undoubtedly contributes to maintenance of some oxidizing potential in the mitochondria during anoxia.

As for mechanisms assuring that coenzyme A (CoASH) supplies be neither accumulated nor depleted, bivalves display at least two possibilities. Firstly, succinic and propionate thiokinases could tumble CoASH between them during the production of propionate. And secondly, the same coupling between ketocarboxylate dehydrogenases and thiokinases as discussed in chapter 2 for helminths is presumed to operate to some extent at least in bivalves.

The energy yield of anaerobic metabolism in marine bivalves is potentially as high as that in helminths. The fermentation of glucose to propionate theoretically yields 6 moles ATP/mole glucose, while aspartate fermentation to succinate yields a mole of ATP/mole aspartate. In the anoxic oyster heart the ATP yield is lower: 3 moles ATP/mole of glycogen-derived glucose plus 1 mole ATP/mole of aspartate. Hence, by comparison with classical glycolysis, the energetic efficiency in anoxic bivalves is increased substantially, as is metabolic versatility.

Probably the most important consequence of the improved versatility in bivalves is the capability of generating a larger number of coupling metabolites required for other metabolic functions. In contrast to the four such coupling metabolites that can be generated by glycolysis, the anaerobic metabolism of bivalves can generate at least eight of the ten required components. As in helminths, it is this ability that perhaps more than any other underpins the long-term anaerobic capacity of marine bivalves.

Suggested Readings

COLLICUTT, J. M., and HOCHACHKA, P. W. 1977. The anaerobic oyster heart: coupling of glucose and aspartate fermentation. *J. Comp. Physiol.* 115:147–157.

DE ZWAAN, A., KLUYTMANS, J. H. F. M., and ZANDEE, D. I. 1976. Facultative anaerobiosis in molluscs. *Biochem. Soc. Symp.* 41:133–168.

FIELDS, J. H. A., HOCHACHKA, P. W., ENG, A. K., RAMSDEN, W. D., and WEINSTEIN, B. 1979. Alanopine and strombine are novel imino acids produced by a dehydrogenase found in the adductor muscle of the oyster, *Crassostrea gigas*. *Arch. Biochem. Biophys.*, in press.

HOCHACHKA, P. W. 1976. Design of metabolic and enzymic machinery to fit lifestyle and environment. *Biochem. Soc. Symp.* 41:3–31.

Chapter Four

Coupled Glucose and Arginine Metabolism in Cephalopods

Thus far we have seen at least two ways in which glucose and amino acid metabolism can be linked. One way involves redox coupling of mitochondrial reactions, a specific example being the interaction between keto carboxylate dehydrogenases in leucine catabolism to isovalerate, and fumarate reductase in glucose fermentation to succinate. This organization is well typified by some helminths, for example, *Fasciola.* A second way involves a redox coupling in the cytosol, a specific example being the interaction between cytoplasmic malate dehydrogenase (MDH) in aspartate fermentation to succinate and glyceraldehyde 3-phosphate dehydrogenase (GAPDH) in glucose fermentation to alanine. This interaction is well exemplified by the anoxic oyster heart, and is further promoted by an interaction between aspartate and alanine aminotransferases: glutamate, the product of the former reaction, is the substrate for the latter.

Cephalopods, in contrast, utilize a direct chemical coupling of glucose and amino acid fermentation to balance redox in the cytoplasm. This is because energy metabolism in cephalopod muscles differs somewhat from that in vertebrates since arginine phosphate replaces creatine phosphate as the phosphagen, and lactate dehydrogenase (LDH) is either absent or low in activity. The terminal step in anaerobic glycolysis is taken over by octopine dehydrogenase (ODH) catalyzing the reaction:

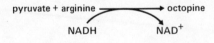

Octopine consequently appears instead of lactate as the primary anaerobic end product (fig. 4.1). Although the role of octopine formation and its subsequent fate is still not fully understood, a lot of information is already available on octopine dehydrogenase—a rather novel situation where we find we know more about the enzyme than the metabolism of its product, octopine.

Because ODH and LDH play similar metabolic roles, it may be expected that they are enzymatically similar. The two enzymes do indeed show some similarities. Thus the subunits of both enzymes are of similar size and have a blocked amino terminus. The active sites of both LDH and ODH contain a catalytically important histidine residue. Both enzymes have essential cysteine residues. And finally, both mammalian M_4 LDH and molluscan ODH contain one cysteine which is more reactive than the essential one.

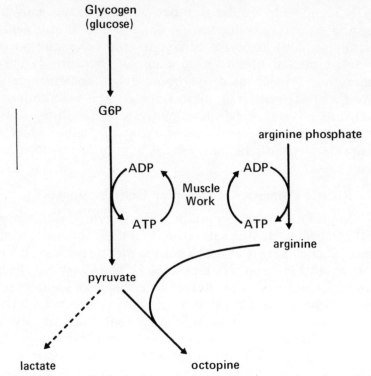

Figure 4.1 Overall scheme of anaerobic glucose metabolism in cephalopod muscles.

However, despite these similarities, the enzymes differ in several important ways. One of their most important differences is in the types of bond systems at which reduction occurs. In LDH, as in most dehydrogenases, a carbon–oxygen bond is the reactive species, while in ODH, as in glutamate dehydrogenase, a carbon–nitrogen bond system fills this role. The analogy between GDH and ODH also includes the binding and catalytic functions of the coenzyme. In both of these enzymes, the hydride transfer is B-side specific, in contrast to the A-side specificity of LDH. Furthermore, the spectral properties of the ODH/NADH binary complex resemble those of the binary glutamate dehydrogenase/NADH complex, but differ from those of the LDH/NADH complex. Fluorometric data also indicate similar conformational changes occurring on substrate binding in glutamate and octopine dehydrogenases.

In most, if not all, of these kinds of characteristics ODH is probably similar to other dehydrogenases catalyzing the reductive condensation of keto and amino acids. Alanopine dehydrogenase catalyzing the condensation of pyruvate plus alanine to form alanopine has already been mentioned. Analogous dehydrogenases are known to catalyze the condensation of: pyruvate + glycine → strombine; pyruvate + lysine → lysopine; pyruvate + histidine → histopine; pyruvate + ornithine → oxtopinic acid; 2-KGA + lysine → saccharopine; and 2-KGA + arginine → nopaline (see chapter 1).

Kinetic Properties of Octopine Dehydrogenases

In all ODHs thus far studied, the enzyme follows normal Michaelis-Menten saturation kinetics. The most complete kinetic data are available for the mantle muscle ODHs from squid and octopus (tables 4.1-4.4). The K_m for NADH is similar in both and is unaffected by either arginine or pyruvate. No substrate inhibition by arginine is detected; on the other hand, pyruvate is inhibitory but only at levels (10 mM) that are far above the physiological range.

The K_m for NAD^+ is slightly lower in the octopus than in the squid, and that for octopine is markedly lower (tables 4.1 and 4.2). Also, the K_m increases if the coenzyme concen-

Table 4.1 Octopus mantle muscle octopine dehydrogenase: apparent K_m values for substrates.

Substrate	Cosubstrate	K_m, mM
NADH	6 mM arginine, 1.5 mM pyruvate	0.02
Arginine	0.2 mM NADH, 1.5 mM pyruvate	7.1
Pyruvate	0.2 mM NADH, 6 mM arginine	2.9
NAD$^+$	0.7 mM octopine	0.13
Octopine	1 mM NAD$^+$	0.8

Source: Data from Fields, Baldwin, and Hochachka (1976), with modification.

tration is decreased. This suggests either that the octopus enzyme can function more easily in both directions under physiological conditions, or that octopine concentrations reach higher levels in the squid. However, octopine concentrations in mantle muscle of these species have not yet been determined; therefore, this question cannot be resolved at this time.

Octopine gives a mixed inhibition pattern with both arginine and pyruvate; however, it is more inhibitory with respect to arginine. The K_i values were similar for both species (tables 4.3 and 4.4), but it is difficult to attach any physiological importance to this in terms of regulation of activity, because the levels of octopine in the tissue are unknown.

Pyruvate gives a mixed pattern of inhibition with re-

Table 4.2 Squid mantle muscle octopine dehydrogenase: apparent K_m values for substrates.

Substrate	Cosubstrate	K_m, mM
NADH	6 mM arginine, 1.2 mM pyruvate	0.03
Arginine	0.2 mM NADH, 0.24 mM pyruvate	13.5
Pyruvate	0.2 mM NADH, 3 mM arginine	4.0
NAD$^+$	10 mM octopine	0.15
Octopine	1.0 mM NAD$^+$	4.4

Source: Data from Fields, Baldwin, and Hochachka (1976), with modification.

Table 4.3 Octopus mantle muscle octopine dehydrogenase: inhibition constants (K_i values) for various inhibitors.

Substrate	Inhibitor	K_i, mM
NADH	NAD$^+$	0.16
	ATP	1.9
	ADP	0.8
	AMP	4.2
Pyruvate	Octopine	14.3
Arginine	Octopine	6.4
NAD$^+$	NADH	0.045
	ATP	4.8
	ADP	6.2
	AMP	16
Octopine	Pyruvate	9
	Arginine	5

Source: Data from Fields, Baldwin, and Hochachka (1976), with modification.

spect to octopine, but the K_i values are very high; therefore this probably has little physiological relevance. Arginine inhibition is mixed in the squid, but in the octopus it is complex, apparently tending toward an uncompetitive pattern. The K_i is about 5 mM, and this is certainly in the physiolog-

Table 4.4 Squid mantle muscle octopine dehydrogenase: K_i values for various inhibitors.

Substrate	Inhibitor	K_i, mM
NADH	NAD$^+$	0.4
	ATP	2.7
	ADP	2.8
	AMP	3.9
Pyruvate	Octopine	12
Arginine	Octopine	10
NAD$^+$	NADH	0.012
Octopine	Pyruvate	19.0
	Arginine	5.4

Source: Data from Fields, Baldwin, and Hochachka (1976), with modification.

ical range for arginine. Both enzymes show low K_i values for NAD$^+$ and NADH (tables 4.3 and 4.4). Of a series of other metabolites tested, only the adenylates, being competitive inhibitors of NADH, are effective; aspartate, glutamate, taurine, proline, alanine, glycine, arginine phosphate, α-glycerophosphate, fructose biphosphate, phosphoenolpyruvate, dihydroxyacetone phosphate, 2-ketoglutarate, citrate, succinate, and malate have no effect.

Although any or all of the catalytic properties described may contribute to in vivo ODH regulation, they lack the

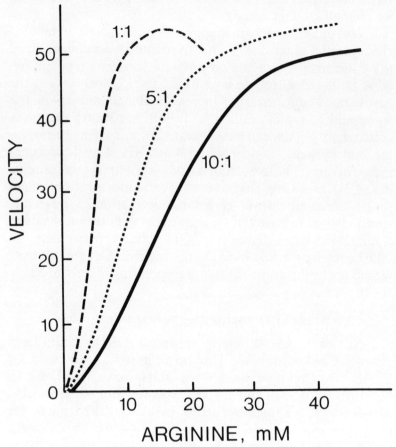

Figure 4.2 Squid mantle octopine dehydrogenase. Effect of covarying arginine and pyruvate concentrations while keeping the arginine: pyruvate ratio constant. Data are plotted with respect to the arginine concentration, varying arginine: pyruvate ratio.

specificity needed for versatile control of the enzyme through various metabolic transitions. Two features, however, are known that may contribute specifically to turning the enzyme on or off at appropriate times. The first involves pH; the second involves interacting effects of pyruvate and arginine.

Like all ODHs, in squid and octopus mantle both enzymes for the forward (octopine-forming) direction show a relatively sharp pH optimum somewhat below pH 7. If mantle muscle pH varies systematically with different activity states, it clearly could supply the system with a signal for changing ODH activity.

The second mechanism appears much more metabolically specific, since the K_m for pyruvate is dependent upon the concentration of the cosubstrate (arginine) and the reverse is likewise true. In both cases, the K_m decreases as the cosubstrate concentration increases; hence, enzyme activity would be rapidly enhanced if both substrates increased concurrently. This can be illustrated by covarying the arginine and pyruvate concentrations at a fixed ratio of arginine:pyruvate. The resulting saturation curve is sigmoidal (fig. 4.2), indicating that over critical concentration ranges to be expected during glycolytic activation, a relatively small change in pyruvate + arginine concentration leads to a relatively large change in ODH activity; the change in ODH activity for this kind of enzyme is in fact greater than would occur if the enzyme were responding to only one substrate.

Two Kinds of Octopine Dehydrogenases

At least two different types of ODHs have been characterized which we can term low K_m and high K_m ODHs. The first type, the low K_m ODHs, have been found in *Nautilus* (table 4.5) and scallop muscles, and these are also low-activity enzymes occurring in about 40-80 units per gram per min at 25°C. The second type, or the high K_m ODHs, have been found in squid muscles and in octopus muscles (tables 4.1-4.2). Particularly in the more sluggish squids and the octopus, these are high activity enzymes, the octopus activity getting as high as 600 units per gram

Table 4.5 *Nautilus* spadix muscle octopine dehydrogenase. Increasing availability of one substrate increases the enzyme affinity for the cosubstrate.

Substrate	Cosubstrate, mM	K_m, mM
Pyruvate	1.5	1.3
	6.0	0.7
	30.0	0.3
Arginine	0.45	7.7
	0.9	5.3
	3.0	3.7

Source: Data from Hochachka et al. (1977), with modification.

per min at 25°C. The absolute value for the kinetic constants shown in tables 4.1, 4.2, and 4.5 indicate that enzyme affinities for arginine and NADH are probably within the physiological range for both the squid or octopus ODH and the *Nautilus*-type ODH. However, enzyme affinity for pyruvate in the case of squid mantle or octopus mantle is extremely low, K_m values for pyruvate being in the 2-4 mM range, whereas in the case of the low K_m type ODHs from *Nautilus*, these are in the 0.3 to 0.5 mM range. The implications are that octopine formation in squid and octopus muscles may be favored only under extreme conditions when pyruvate concentrations get very high. And second, the formation of octopine should be more readily demonstrable in the muscles of *Nautilus* and scallop.

Experimentally, these implications have, in fact, been verified, and it is a striking observation that in the case of *Loligo*, at least in captivity, octopine concentrations in the mantle muscle cannot be shown to accumulate above a few μmol per gram wet weight, while in the *Nautilus* retractor, concentrations as high as 25 to 30 μmol per gram wet weight are readily detected.

Functional Significance of Octopine Dehydrogenase

Three functions of ODH seem evident. The first of these is to serve in redox balance in a manner formally analogous to lactate dehydrogenase. Second, octopine formed by the ODH reaction can serve as an arginine sink. Under some conditions this may be of advantage, since the pK of arginine is high. The compound is so basic it will absorb atmospheric CO_2 into solution. The total arginine phosphate/arginine pool can be as high as 70 μmol per gram, and clearly any significant changes in arginine concentration must either be potently buffered or prevented. One obvious way of doing this is to convert it to octopine. By comparison with arginine, octopine is a very mild acid, with an isoelectric point of about 5.9. Third, free arginine may not only be detrimental to intracellular pH regulation, but may also have devastating nonspecific effects on the catalytic properties of enzymes. Hence it is very much in the interest of the cell to closely regulate the concentrations of free arginine.

Isozymes of Octopine Dehydrogenase

Although the production of octopine during anaerobic muscle work in cephalopods has been adequately established, its subsequent metabolic fate has not been clarified. Since ODH catalyzes a fully reversible reaction, some portion of the octopine formed during anaerobic metabolism could be catabolized in situ during recovery at sites of formation. However, some octopine could also be released into the blood for further metabolism elsewhere, a more complicated situation calling for ODH function either in octopine formation or octopine oxidation. In this connection it is particularly instructive that octopine dehydrogenase, like LDH, has been found to occur in tissue-specific isozymic form. In *Nautilus* muscles two such forms have been identified. Similarly, in the pelagic squid, *Symplectoteuthis*, at least two forms of ODH are known. In *Sepia* at least three, and probably four, ODH isozyme forms have been found. As far as is known, all ODHs occur as single subunit enzymes of approximately 35,000 molecular weight, hence the occurrence of more than one isozyme of ODH within a single individual implies the occurrence of more than one gene for the enzyme. This situation is similar to that of the LDH isozymes, where two genes code for two subunits, which are randomly associated to form five-tetramer LDH isozymes.

Despite compositional differences, interesting parallels have been drawn between ODH and LDH isozyme func tions. In the case of LDHs the muscle-type enzyme is thought to function primarily in lactate production during anaerobic glycolysis, while the heart-type LDH is thought to scavenge lactate released into the blood, and to oxidize it to pyruvate for further metabolism. Functionally, the ODH isozymes seem to be similarly specialized kinetically. Thus the properties of muscle-type ODHs appear to favor function in octopine production during anaerobic glycolysis. In contrast, in some tissues, notably the gills, heart, and brain, electrophoretically fast forms of ODH seem kinetically adapted to function in octopine oxidation (tables 4.6 and 4.7). Interestingly, the muscle-type ODH isozymes seem to occur in very high activities by comparison with the heart-type ODHs. What is more, muscle and heart-type ODHs show similar tissue specificity to muscle and heart-type LDHs.

The above enzyme studies imply the potential for vigorous cycling of glucose, octopine, and arginine through the cephalopod body, but this possibility has only been investigated once and only in one species: the cuttlefish, *Sepia*. The key experimental observations are these:

(1) When *Sepia* is made hypoxic, blood octopine levels rise because of washout from various tissues, with the concomitant drop in blood glucose levels. About 25% as much octopine appears in the blood as would be expected if all the glucose used were fermented.

(2) On return to normoxic conditions, blood glucose

Table 4.6 Squid brain octopine dehydrogenase: apparent K_m values for substrates.

Substrate	Cosubstrate	K_m, mM
NADH	5 mM arginine, 1.2 mM pyruvate	0.01
Pyruvate	0.2 mM NADH, 5 mM arginine	1.3
Arginine	0.2 mM NADH, 1.2 mM pyruvate	4.2
NAD$^+$	0.5 mM octopine	0.1
Octopine	1 mM NAD$^+$	0.2

Source: Data from Fields, Guderley, Storey, and Hochachka (1976), with modification.

Table 4.7 Squid brain octopine dehydrogenase: effect of inhibitors.

Substrate	Inhibitor	K_i, mM
NADH	NAD$^+$	0.04
	ATP	2.5
	ADP	2.0
	AMP	4.2
Arginine	Octopine	2.5
Pyruvate	Octopine	7.2
NAD$^+$	NADH	0.009
Octopine	Pyruvate	6.8
	Arginine	9.0

Source: Data from Fields, Guderley, Storey, and Hochachka (1976), with modification.

concentrations rise to, and in fact overshoot, initial levels with a concurrent fall in blood octopine concentration.

(3) Most of a high dose of octopine infused into *Sepia* blood is metabolized within an hour at 15°C; the disappearance of blood octopine occurs simultaneously with an increase in blood glucose levels.

(4) Similarly, artificially elevated blood glucose levels are returned to normal within an hour, with simultaneous if small pulses in blood octopine and lactate levels.

(5) Artificially elevated blood lactate levels also are returned to normal with simultaneous pulses of octopine and glucose.

(6) Infusion of a bolus of arginine leads to initially high levels (7 μmol/ml), but these are quickly depleted. Dropping arginine levels are reflected in a small and simultaneous rise in blood octopine concentrations.

(7) Several tissues (brain, ventricle, branchial heart, and gill) rapidly concentrate ^{14}C-octopine infused into the organism via the vena cava, but the same process occurs very slowly in mantle muscle.

Chromatographic analysis indicates that over a 20-minute period none of the mantle muscle ^{14}C-octopine is oxidized, while in the gill, ventricle, and brain, 6%, 32%, and 40%, respectively, of the ^{14}C-octopine taken up is converted to arginine + pyruvate.

The main point to be derived from such data is the patent interrelationship of the four metabolites (octopine, arginine, glucose, and lactate) in the blood and by extension in the tissues as well. The simplest interpretation, in need of more vigorous proof in *Sepia* and other cephalopods as well, is that glucose and octopine are related as substrate and end product in anaerobic glycolysis while, through a modified Cori cycle, the pyruvate moiety of octopine and lactate, when added exogenously, is reconverted to glucose via gluconeogenesis. Gluconeogenesis in a tissue (or tissues) different from the ones producing octopine also requires an effective system for recycling arginine between tissues.

Isozymes of Lactate Dehydrogenase

Although the metabolism of lactate is consistent with the above interpretation, the ease with which *Sepia* can metabolize lactate is in a sense surprising. This is because in earlier studies of cephalopod octopine dehydrogenases, it was proposed that ODH and LDH are mutually exclusive: most species showing one were thought not to express the other. This impression arose because insufficient numbers of tissues were examined. Cephalopod muscle typically has very high ODH activity, nearly 100 times the LDH activity in *Sepia*, for example, and from such tissues the impression could easily be obtained of a near absence of LDH but ample ODH activity. However, in some organs, *Sepia* gill for example, ODH and LDH activities are, in fact, similar. In many organs, the ODH:LDH ratio is about 5:10. That is, both dehydrogenases clearly can occur simultaneously, and, as Storey (1977) has shown, the muscle-versus heart-type specializations of one are mimicked by the muscle- versus heart-type specializations of the other. But why a two-dehydrogenase systems?

Double (Octopine and Lactate) Dehydrogenase Systems

The functions of two kinds of dehydrogenases depend upon the metabolic situation. In organs such as the gills the kinetic characteristics of the heart-type ODHs and LDHs favor the oxidation of their respective substrates, octopine and lactate. In other organs, particularly with muscle type

ODH, the occurrence of low LDH activities appears to indicate an auxiliary mechanism for NADH oxidation under anaerobic conditions.

In this view, the three enzymes GADPH, ODH, and LDH together are required to maintain NADH redox balance during high anaerobic glycolytic rates. NAD^+ reduction by GAPDH must be coupled stoichiometrically to NADH oxidation by other dehydrogenases. ODH is the principal catalyst in this coenzyme cycling process. However, ODH, like LDH, is presumed to operate near equilibrium under steady-state conditions, and therefore both octopine and pyruvate must accumulate during anaerobic glycolysis, probably in a ratio of about 30:1. To maintain the NAD^+ redox state under these conditions, an amount of NADH equivalent to the accumulated pyruvate must be oxidized by additional dehydrogenases. In highly anaerobic cephalopod muscles, this is where LDH may come into play, serving as an NADH oxidation mechanism supplementary to ODH function. An entirely analogous system has been described in chicken breast muscle where LDH function is supplemented by α-GPDH function. Indeed, the metabolic advantage (or, in highly anaerobic muscles, even necessity) for such functional interactions among more than one cytoplasmic dehydrogenase may have selected for their development in parallel as coupled metabolic units. Otherwise it would be difficult to understand why they appear over and over again in vertebrate as well as in invertebrate systems.

Double (Malate and Alanopine) Dehydrogenase Systems

The above analysis holds only when both of the dehydrogenase couples produce true anaerobic end products, and does not apply to the malate dehydrogenase/alanopine dehydrogenase couple in the anoxic oyster heart. This is because MDH function, serving as the primary mechanism for cytosolic redox balance, does not lead to a large malate accumulation. In this case, therefore, the usefulness of MDH to redox regulation is not limited by a mass action effect on an enzyme at, or near to, equilibrium. But MDH can become limiting if aspartate pools are depleted. This, too, is a condition that calls for a double dehydrogenase system, for under these conditions a redox control mechanism auxiliary to MDH must be available if anaerobic glycolysis is to continue. Alanopine dehydrogenase is a good solution to this problem since alanine concentrations rise as aspartate levels fall in the anoxic oyster heart and thus alanine serves as a strategic metabolic signal for turning on a supplementary NADH oxidation system: it signals when alanopine dehydrogenase should be turned on (as MDH runs out of substrate because of aspartate depletion) and by how much it should be turned on (to levels high enough to prevent further alanine accumulation). The role of MDH in cephalopods, however, is not quite the same.

Double (Malate and Octopine) Dehydrogenase Systems

In cephalopods, an MDH/ODH double dehydrogenase system may take on importance in highly anaerobic tissues displaying high levels of both enzymes. The spadix muscle of *Nautilus* supplies a case in point. The ultrastructure of spadix muscle indicates a very low oxidative capacity (mitochondria occupy only about 0.6% of the fiber area). Not surprisingly, a mitochondrial marker enzyme, such as citrate synthase, occurs at extremely low activities, while ODH and MDH occur at about 40 and 80 units/gm, respectively. Since there is a dearth of mitochondria, the bulk of the MDH activity must be cytoplasmic. Aspartate occurs at about 3-4 μmol/gm, and could readily be transaminated to oxaloacetate (OXA) to act as an hydrogen acceptor for NADH oxidation.

It is probable that in initial stages of anaerobic muscle activation, arginine phosphate hydrolysis supports muscle work, leading to increasing availability of arginine. But to "dump" this arginine into octopine also requires an increased supply of pyruvate from anaerobic glycolysis, and it is tempting to suggest that under these conditions MDH serves to oxidize glycolytically formed NADH, and thus sets the stage for pyruvate formation and an ODH flare-up.

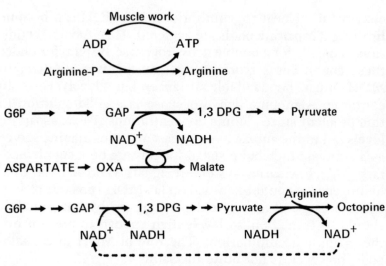

Figure 4.3 Proposed anaerobic activation sequence in *Nautilus* spadix.

Such a view implies that MDH in the *Nautilus* spadix is initially the primary redox regulation mechanism (fig. 4.3). That would explain why molluscan MDHs typically display higher affinities for OXA than do ODHs for pyruvate and arginine. Since the supplies of aspartate are not large and since aspartate mobilization to malate involves two equilibrium enzymes, the amount of malate accumulating must be small. However, it would not have to be large. The $K_{m(pyruvate)}$ for spadix ODH is about 1 mM at low arginine and drops to about 0.3 mM as arginine levels rise. Basal levels of pyruvate are probably in the 0.05 to 0.1 mM range. That means the total amount of malate accumulating need increase the malate pool by only about 0.2 μmol/gm at minimum, about 1 μmol/gm at maximum, and these magnitudes are well within the capabilities of the aspartate aminotransferase and MDH reaction sequence. The net effect of these interactions is an MDH-mediated priming of anaerobic glycolysis to increase the availability of pyruvate. A similar role for MDH in the mammalian heart may explain why during initial stages of anaerobic glycolysis MDH, not LDH, serves as the primary mechanism for oxidizing glycolytically formed NADH.

Hypoxia Adaptations in Cephalopods

From this brief analysis it should be clear that anaerobic metabolism in cephalopods differs in at least three ways from vertebrate glycolysis. First, arginine phosphate replaces creatine phosphate as the phosphagen. Second, and perhaps correlated with this feature, octopine dehydrogenase becomes the main mechanism for oxidizing glycolytically formed NADH at least in some organs (such as muscle); octopine, as a result, replaces lactate as the chief anaerobic end product formed. Third, despite the occurrence of ODH, LDH is represented, probably in all tissues; in some, such as the gill, LDH activity actually exceeds ODH activity.

Nevertheless, in many fundamental ways, cephalopod anaerobic glycolysis shows startling parallels with that in vertebrates. Thus, the fuel source in both appears to be primarily carbohydrate. Redox balance does not require participation of amino acid catabolism. The energy yield is 2 moles ATP/mole glucose fermented. And finally, the system generates only 4 of the requisite 10 coupling intermediates needed for sustained anoxia (table 4.8). In all these fundamental characteristics, anaerobic metabolism in cephalopods seems far less versatile and indeed less well developed for sustained function than in bivalve molluscs or helminths. Thus it is not surprising that as a group cephalopods are notably O_2-dependent organisms. Although unusually anoxia-tolerant species may yet be discovered, we would expect this tolerance to be based on principles different from those used by bivalves but perhaps similar to those

Table 4.8 Hypoxia adaptations in cephalopods.

Requirement	Solution
Storage energy	Carbohydrate
Redox regulation	Octopine dehydrogenase + lactate dehydrogenase
ATP yield	
Glucose → octopine	2
Coupling metabolites	4

utilized by vertebrates (for example, involving O_2 conservation mechanisms, interorgan metabolic cooperation, and so forth).

The usefulness of the cephalopods as experimental organisms in the study of hypoxia tolerance derives from their similarities with vertebrates: almost identical solutions to the essential requirements of anaerobic metabolism are utilized by both groups despite a number of important enzymatic differences. This raises the intriguing possibility of testing the functional significance of specific mechanisms in vertebrate glycolysis by searching for parallel phenomena in cephalopod glycolysis. In this way, interpretations of one system can be cross-checked by analysis of the other, as is in fact illustrated in the case of double dehydrogenase systems for the regulation of cytoplasmic redox potential.

Suggested Readings

BIELLMANN, J. F., BRANLANT, G., and OLOMUCKI, A. 1973. Stereochemistry of the hydrogen transfer to the coenzyme by octopine dehydrogenase. *FEBS Lett.* 32:256–256.

FIELDS, J. H. A., BALDWIN, J., and HOCHACHKA, P. W. 1976. On the role of octopine dehydrogenase in cephalopod mantle muscle metabolism. *Can. J. Zool.* 54:871–878.

FIELDS, J. H. A., GUDERLEY, H., STOREY, K. B., and HOCHACHKA, P. W. 1976. The pyruvate branchpoint in squid brain: competition between octopine dehydrogenase and lactate dehydrogenase. *Can. J. Zool.* 54:879–885.

GRIESHABER, M., and GADE, G. 1976. The biological role of octopine in the squid, *Loligo vulgaris* (Lamarck). *J. Comp. Physiol.* 108:225–232.

———. 1977. Energy supply and the formation of octopine in the adductor muscle of the scallop, *Pecten jacobaeus* (Lamarck). *Comp. Biochem. Physiol.* (B) 58:249–252.

HOCHACHKA, P. W., FRENCH, C. J., and MEREDITH, J. 1978. Metabolic and ultrastructural organization in *Nautilus* muscles. *J. Exp. Zool.* 205:51–62.

HOCHACHKA, P. W., HARTLINE, P. H., and FIELDS, J. H. A. 1977. Octopine as an end product of anaerobic glycolysis in the chambered nautilus. *Science* 195:72–74.

OLOMUCKI, A., HUC, C., LEFEBURE, F., and VAN THOAI, N. 1972. Octopine dehydrogenase: evidence for a single-chain structure. *Eur. J. Biochem.* 28:261–268.

PHO, D. B., OLOMUCKI, A., HUC, C., and VAN THOAI, N. 1970. Spectrophotometric studies of binary and ternary complexes of octopine dehydrogenase. *Biochim. Biophys. Acta* 206:46–53.

STOREY, K. B. 1977. Tissue specific isozymes of octopine dehydrogenase in the cuttlefish, *Sepia officinalis:* the roles of octopine dehydrogenase and lactate dehydrogenase in *Sepia. J. Comp. Physiol.* 115:159–169.

STOREY, K. B., and STOREY, J. M. 1979. Octopine metabolism in the Cuttlefish, *Sepia officinalis:* octopine production by muscle and its role as an aerobic substrate for non-muscular tissues. *J. Comp. Physiol.* (B) 131:311–320.

STOREY, K. B., STOREY, J. M., JOHANSEN, K., and HOCHACHKA, P. W. 1979. Octopine metabolism in *Sepia officinalis:* effect of hypoxia and metabolite loads on the blood levels of octopine and related compounds. *Can. J. Zool.,* in press.

Chapter Five

Key Elements of Anaerobic Glycolysis

It should now be evident that in all cells able to sustain significant periods of anoxia, some provision is made for satisfying several fundamental metabolic requirements. In invertebrates, exploitative strategies of biochemical adaptation allow the development of metabolic systems that are surprisingly efficient in supplying most of these needs. Vertebrate organisms, in contrast, rely almost exclusively upon anaerobic glycolysis for these needs (fig. 5.1). In this process, glycogen, the storage form of energy, is catabolized to the level of lactate during anaerobiosis. Redox balance is maintained by a functional $1:1$ integration of glyceraldehyde 3-phosphate dehydrogenase (GAPDH) and lactate dehydrogenase (LDH), catalyzing oxidative and reductive steps, respectively. Useful energy production in the form of ATP occurs at the phosphoglycerate kinase and pyruvate kinase reaction steps, a net of 2 moles of ATP being formed per mole of glucose. It is useful to review this overall system, to better understand how it can be maximized or at least improved, to support work under stressful hypoxic conditions.

The simplest way to improve glycolytic capacity is to increase either the amount of substrate (glycogen) stored or the amount of glycolytic enzymes. All else being equal, more substrate stored plus a greater enzymatic potential for fermenting it allows for improved tolerance of anoxia. Tuna white muscle, for example, is capable of perhaps the most

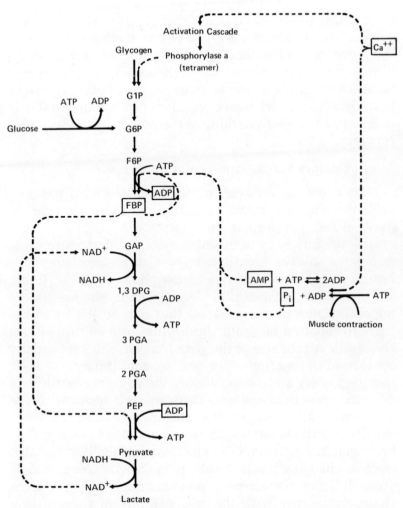

Figure 5.1 Scheme of anaerobic glycogenolysis in vertebrate muscle illustrating how several key regulatory modulators (boxed) lead to integration of muscle contraction with metabolic activation.

intense anaerobic glycolysis known in nature. Enzymes such as lactate dehydrogenase occur at over ten times the level found in other vertebrate muscles. Associated with such high enzyme levels are glycogen storage depots that are substantially higher than in other teleost white muscles. Mammalian skeletal muscle, similarly, may at times rely heavily upon anaerobic glycolysis as the sole means for producing energy; to this end, ample glycogen is stored in mus-

cle and this tissue maintains higher levels of glycolytic enzymes than are found anywhere else in the body. However,
the properties of metabolic processes (including glycolysis)
ultimately reside in the nature of their component enzymes
rather than simply in the amount present; hence, to assess
how glycolysis in whatever organism can be improved it is
necessary to know something of the properties of glycolytic
enzymes.

Regulatory versus Nonregulatory Enzymes

There are two kinds of enzyme functions readily identifiable in glycolysis: enzyme functions far from, and near to,
equilibrium. The former are catalyzed by regulatory enzymes; the latter, by nonregulatory ones. Metabolite measurements always identify at least three regulatory enzymes: glycogen phosphorylase, phosphofructokinase, (PFK),
and pyruvate kinase (PK). And where glucose is an
important alternative substrate, hexokinase (HK) also appears to function far from equilibrium as a control site in
glycolysis. All the rest of the glycolytic enzymes are usually
considered to function at, or near to, equilibrium.

Regulatory and nonregulatory enzymes play strikingly
different roles in glycolysis. Thus, in each segment of the
glycolytic path, one of the above regulatory enzymes
usually contributes to setting the overall rate of carbon flow
by responding sensitively to changing metabolite "signals"
(such as changing levels of substrate, cofactors, coenzymes,
or modulators). In contrast, nonregulatory enzymes seem
charged primarily with the responsibility of transmitting
carbon along the pathway, with utmost speed. Not surprisingly, regulatory and nonregulatory enzymes seem to be selected for contrasting properties; that is, control versus high
rate. As a result, they do not work quite the same way.

The most fundamental features of all enzymes are binding and catalysis. Although these are conceptually separable
functions, in the "real world" they are related. Haldane was
probably first to raise the possibility of that relationship. He
suggested in 1930 that binding energies may be used to distort the substrate to the structure of the products, an idea
that has served as the point of departure for much subse-

quent exploration of the various ways in which the binding energy of the enzyme and substrate may be used to lower the activation energy of the chemical steps. As a result a relatively simple picture has emerged. It indicates that the maximum binding energy between an enzyme and a substrate occurs when each binding group on the substrate is matched by a binding site on the enzyme. In this case the enzyme is said to be complementary in structure to the substrate. Since the structure of the substrate changes throughout the reaction, becoming first the transition state and then the products, the structure of the unaltered binding site can be complementary to only one form of the substrate. Fersht (1977) has argued that it is catalytically advantageous for the enzymes to be complementary to the structure of the transition state of the substrates rather than the original structures. If this happens, the increase in binding energy as the structure changes to that of the transition state lowers the activation energy. This situation may apply to some nonregulatory enzymes in glycolysis.

Conversely, in regulatory enzymes where a high substrate affinity, adjustable by enzyme modulator interactions, is more advantageous than high catalytic rate per se, selective pressure may favor enzymes whose binding regions are complementary to the unaltered substrate.

There are important consequences stemming from an enzyme being complementary to the transition state rather than to the initial substrate: (1) Activation energy is lower, as already implied. (2) The ratio, k_{cat}/K_m, a secondary rate constant which determines velocity at low substrate levels, is higher (in fact is maximized when k_{cat} approaches a diffusion-controlled limit).* (3) K_m is high (that is, substrate binding is weak because of strong binding of the transition state). Some investigators have argued that this feature is selected because the maximum reaction rate for a particular substrate level depends on the individual values of k_{cat} and K_m. If k_{cat}/K_m is held constant, higher reaction rates are obtained when K_m is greater than [S], the estimated physiolog-

---

* $k_{cat} = V_{max}/$enzyme concentration. K_m = Michaelis constant, in units of substrate concentration.

ical substrate level. Thus the maximization of rate requires high values of K_m. This concept of maximization of K_m at constant k_{cat}/K_m in fact contradicts widely held beliefs that strong bindings, or low K_m, is an important component of metabolic enzymes. In contrast, the hypothesis predicts that equilibrium enzymes should evolve to bind substrate relatively weakly.

In theory, opposite consequences would be expected from an enzyme being complementary to the undistorted initial substrate rather than to the transition state; although this situation may be typified in some regulatory enzymes, not enough evidence is available to draw unequivocal conclusions. That evidence which is in indicates quite a different picture from nonregulatory enzymes. For example, in a series of eight genetic variants of erythrocyte PK, a regulatory enzyme subject to strong activation by fructose biphosphate, only one has a K_m value lower than normal, while in all the others, the K_m rises (table 5.1). The V_{max}/K_m ratio (proportional to k_{cat}/K_m) similarly follows no simple pattern: One variant shows a five-fold drop. Four variants show about a two-fold increase in the V_{max}/K_m ratio, while the K_m

Table 5.1 Biochemical characteristics of eight pyruvate kinase variants and a classical type pyruvate kinase deficiency.

Enzyme variant	K_m for PEP	V_{max}
Classical type PK deficiency	1.4	2.1
PK "Kiyose"	3.2	14.5
PK "Tokyo I"	4.0	13.0
PK "Nagasaki"	5.2	10.9
PK "Maebashi"	5.8	1.7
PK "Sapporo"	3.2	26.8
PK "Tsukiji"	4.5	32.8
PK "Tokyo II"	1.0	2.9
PK "Ube"	1.7	15.9
Normal range	1.4	14.7

Source: Miwa et al. (1975), with modification.

values vary from 0.7 to four-fold higher than that of the "normal" PK. Three variants show about a six-fold rise in the V_{max}/K_m ratio while the K_m increases from 1.2 to three-fold. Of the eight variants, only two (from the latter three) regulatory pyruvate kinases would fit the general pattern discussed by Fersht (1977).

In both kinds of enzyme, however, there clearly must be a limit to maximizing rate or to maximizing enzyme affinity for substrate. Thus, for nonregulatory enzymes selected for speed, how far a K_m increases relative to the substrate concentration depends on the change in structure on going from the substrate to the transition state. A limit must eventually be reached when any increase in K_m must be matched by a weakening of transition-state binding. Conversely, for regulatory enzymes, the trend toward high affinity (low K_m) must also reach some compromise limit, when the advantage of high affinity is balanced by the disadvantages of too stable an enzyme-substrate complex or too high an activation energy.

Criteria of Adaptation of Equilibrium Enzymes

According to Fersht's analysis (1977), two criteria can be used to assess the "adaptational state" of an enyzme whose function is to maximize rate: the degree to which k_{cat}/K_m is maximized; and the degree to which K_m is greater than [S]. With respect to the first notion, the maximum value of k_{cat}/K_m is the rate constant for the diffusion-controlled encounter of the enzyme and substrate, and this is about 10^8 to 10^9 S^{-1} M^{-1}. A perfectly developed enzyme should have k_{cat}/K_m in the range of 10^8 to 10^9 S^{-1} M^{-1}. With respect to the second notion, a high adaptational state re-requires a high $K_m/[S]$ ratio.

Do these criteria apply for nonregulatory enzymes in glycolysis? If they do, these two key parameters should both increase when and where glycolytic potential increases. In the vertebrate body, skeletal muscle displays the highest glycolytic potential and it is instructive to compare a number of muscle enzymes with their homologues in other organs. The best information available probably is for LDHs and aldolases.

At the outset, it is important to recall that LDH in mammals is differentiated into two isozyme types. These are tetrameric enzymes, about 140,000 MW, formed through the random association of two kinds of subunits. Aerobic tissues, such as the heart and brain, possess predominantly H_4 LDH, while skeletal muscle (and other tissues displaying relatively anaerobic metabolism) possess predominantly M_4-type LDH. Most tissues in fact synthesize both subunits, in tissue-specific amounts, and hence possess unique distributions of five LDH isozymes: H_4, H_3M_1, H_2M_2, H_1M_3, and M_4. The catalytic properties of these isozymes have been extensively studied; the amino acid sequences of H- and M-type subunits are known, as is the three-dimensional crystal structure.

It is widely accepted that H_4 LDHs have a low K_m for pyruvate and NADH, are sensitive to inhibition by high pyruvate levels and/or high lactate levels, and have a high affinity for lactate. Thus, they are kinetically well suited for function in an aerobic metabolism since they will not compete well for pyruvate (which is "spared" for the Krebs cycle). Moreover, they can compete well for lactate and thus initiate lactate oxidation.

M_4 LDHs have higher K_m values for pyruvate and NADH, are insensitive to pyruvate or lactate inhibition, and have a low affinity for lactate. Thus, they are thought to function as pyruvate reductases in anaerobic glycolysis.

Interestingly enough, k_{cat}/K_m for the muscle enzyme is nearly an order of magnitude higher than in the case of the H_4 LDH (table 5.2). Moreoever, even the highest value is still some two orders of magnitude lower than the value for an enzyme that by this criterion would be perfectly adapted. Thus, as anticipated, M_4 LDH appears to have been specialized to increase reaction rate, which would clearly be useful for transmitting sudden bursts of carbon through the reaction. A similar picture emerges from the aldolase data, the k_{cat}/K_m ratio being highest for the isozyme (A) found in the most glycolytic tissue (muscle). But because M_4 LDH and aldolase A both are still a long way from reaching the maximal rate possible under physiological conditions (that limited by diffusional encounter), we may expect that in other

Table 5.2 Adaptational state of lactate dehydrogenase and aldolase isozymes by k_{cat}/K_m criterion.

Enzyme form	$k_{cat}/K_m S^{-1}M^{-1}$
Beef M_4 LDH	1.8×10^6
Beef H_4 LDH	3.9×10^5
Aldolase A (muscle)	1.7×10^7
Aldolase B (liver)	4.1×10^6
Aldolase C (brain)	5.7×10^6
Theoretically best adapted form	$10^8 - 10^9$

Source: Data for lactate dehydrogenase taken from Borgmann and Moon (1975) and Borgmann et al. (1975) for the forward (pyruvate reductase) reaction at 30°C, with modification. Data for aldolase taken from Penhoet et al. (1969), with modification.

organisms where anaerobic glycolysis may be even more finely tuned than in mammalian muscle, this kinetic characteristic may be even further adjusted. But is the same true for $K_m/[S]$?

$K_m/[S]$ and Isozymes in Glycolysis

To answer this question, it is not only necessary to understand the properties of the isolated enzyme; it is also necessary to know something of the intracellular concentrations of substate. Fortunately, good data are available for both the muscle and heart LDH reactions (table 5.3). These data show that for the resting state, the value of $K_m/[S]$ is some ten- to-twenty-fold higher in muscle than in heart, a result consistent with the above predictions, a high ratio of $K_m/[S]$ being favored in a highly glycolytic tissue. However, the significance of this different between muscle and heart is difficult to appreciate, since it disappears when working states are compared. The latter situation, moreover, seems to be the rule rather than the exception for enzymes in glycolysis.

For most nonregulatory enzymes in glycolysis, k_{cat}/K_m data are not available, but information is available on Michaelis constants and intracellular substrate concentrations for several enzymes and tissues. Those for aldolases A, B,

Table 5.3 Adaptational state of H_4 and M_4 lactate dehydrogenases by $K_m/[S]$ criterion.

Tissue	Pyruvate μmol/gm wet wt	$K_{m(pyruvate)}$ mM	$K_{m(pyruvate)}/$[pyruvate]
Heart (at varying work loads)	0.04	0.07	0.6
Resting muscle	0.04 (lower range)	0.5	13.0
	0.08 (upper range)	0.5	6.5
Working muscle	1	0.5	0.5

Source: Data from Edington et al. (1975), Williamson and Brosnan, (1974), Everse and Kaplan (1975), with modification.

and C in muscle, liver, and brain, respectively, are particularly clear (table 5.4). These data show that the $K_{m(FBP)}$ values for the three enzymes tend to be one to two orders of magnitude lower than estimated tissue substrate concentrations, while the reverse trend appears for glyceraldehyde 3-phos-

Table 5.4 $K_m/[S]$ ratios for aldolases A, B, and C in muscle, liver, and brain, respectively.

Enzyme source	Tissue concentration of metabolite, mM	K_m values, mM	
	FBP	$K_{m(FBP)}$	$K_m/$[FBP]
Muscle (isozyme A)	0.1	0.003	0.03
Liver (isozyme B)	0.01	0.001	0.10
Brain (isozyme C)	0.2	0.003	0.015
	DHAP	$K_{m(DHAP)}$	$K_m/$[DHAP]
Muscle (isozyme A)	0.07	2.0	28
Liver (isozyme B)	0.02	0.4	20
Brain (isozyme C)	0.02	0.3	15
	GAP	$K_{m(GAP)}$	$K_m/$[GAP]
Muscle (isozyme A)	0.1	1.0	10
Liver (isozyme B)	0.02	0.3	15
Brain (isozyme C)	<0.01	0.8	>80

Source: Penhoet et al. (1969), with modification.

phate (GAP) and dihydroxyacetone phosphate (DHAP). Although muscle is the most glycolytic tissue of the three (liver the most glucogenic), the $K_{m(FBP)}$/[FBP] ratio is highest for the liver. Similarly, a comparison of K_m/[S] ratios for several nonregulatory enzymes in glycolysis shows no consistent trend toward high values in muscle.

In trying to appreciate why this should be so, one should recall that Fersht (1977) envisaged (for enzymes whose function is to turn over large amounts of substrate quickly) a direct relationship between k_{cat} and K_m and a high k_{cat}/K_m ratio in order to maximize reaction rates. This conclusion, based on studies of the forward reaction for proteases, clearly is not valid for aldolase A, where (compared to the B and C isozymes) a high k_{cat}/K_m ratio is achieved mainly by increasing k_{cat} while maintaining a very low K_m (table 5.4). This suggests that the relationship between the catalytic and binding constants is not fixed by chemical necessity and may be modulated according to metabolic needs. Although remarkably few studies have probed this problem, it appears that by independently modulating k_{cat} and K_m for the forward and back reactions, enzymes can be adjusted to facilitate function in the physiologically required direction at physiologically required rates. Where speed is selected for, high ratios of k_{cat}/K_m remain advantageous but can be generated by relatively independent modulation of the two constants. Although it is not clear which is the more usual case, it is implicit in the current evidence that both mechanisms contribute to improving the glycolytic capacity in muscle. The glycolytic capability is further potentiated by the occurrence of high absolute amounts of enzyme; indeed, muscle is probably unique in having the highest levels of glycolytic enzymes anywhere in the vertebrate body. Taken together, these characteristics (appropriately high k_{cat}/K_m, and high enzyme amount) are well suited for glycolysis in a tissue where high pulses of carbon must be periodically handled.

Whereas nonregulatory enzymes, such as LDH and aldolase, can clearly be adjusted to cope with a large flare-up in glycolytic rate, these enzymes do not cause such pulses in carbon flux. The latter function involves regulatory enzymes that appear to be selected for tight binding of substrate because control is vested in adjusting enzyme-substrate affinity. As we shall see, the consequences of this strategy are quite different than in the case of nonregulatory enzymes. The four main regulatory enzymes in glycolysis are glycogen phosphorylase, phosphofructokinase (PFK), pyruvate kinase (PK), and hexokinase (HK). Whereas some principles of their control may be similar, individual features (such as "on-off" signals) are so unique they are best discussed one at a time. Also since there is some tissue specificity to control patterns, our discussion will center primarily on control of glycolysis in muscle, periodically contrasting it with other organs.

Strategic Positioning of Glycogen Phosphorylase

As is evident in fig. 5.2, glycogen phosphorylase is metabolically strategically positioned at the first step in the pathways and it initiates glycogen mobilization. Physically, it also appears to be strategically placed onto glycogen granules. From much in vivo and in vitro experimentation, glycogen phosphorylase is known to be under hormonal control (by epinephrine, glucagon, insulin, and glucocorticoids) as well as under metabolite control (by glucose and adenosine monophosphate (AMP)). In addition, Ca^{++} as well as other ions may play important controlling functions. The time course of phosphorylase activation by hormones is of the order of minutes, while calcium activation, which in muscle depends on membrane depolarization and can therefore be thought of as a neuron-dependent activation mechanism, occurs in seconds or less. The calcium activation appears to be of primary importance in certain situations, as for example, in diving vertebrates, when blood flow, and hence hormonal signals to muscle, are largely cut off during the dive but when glycogenolysis may need to be activated (see chapter 9).

Net mobilization of glycogen depends not only upon the activation of glycogen phosphorylase but also on the concomitant inhibition of glycogen synthetase; both of these regulatory enzymes occur in two forms, an active *a* form and an inactive *b* form. The *a* and *b* forms are inter-

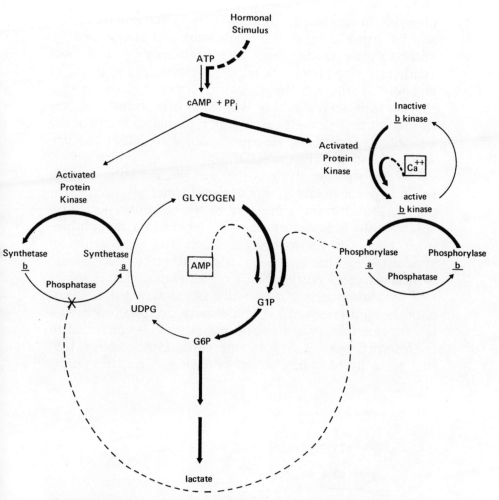

Figure 5.2 Scheme of events during glycogen mobilization stimulated by hormones (for example, epinephrine) or by membrane depolarization and Ca^{++} release. Dark arrows indicate activation; dark crosses indicate inhibition.

convertible by the action of respective kinases and phosphatases (fig. 5.2). Phosphorylase *a* and synthetase *a* functions are mutually exclusive in part because of the inhibitory effect of the former on the production of the latter, and in part because of a regulatory effect of glucose. When glucose concentrations are high (during conditions of glycogen deposition in liver, for instance), glucose directly binds to phosphorylase *a*, leading to inhibition by direct kinetic effects and by stimulating $a \rightarrow b$ conversion. As the percent of phosphorylase *a* decreases, its inhibitory control over synthetase phosphatase is released, allowing a sparking of glycogen synthetase (fig. 5.3). Activation of glycogen synthetase thus serves as a pull mechanism stimulating the flow of glucose into glycogen which decreases levels of the immediate precursors, glucose 6-phosphate (G6P) and uridine diphosphate glucose (UDPG).

When glycogenolysis is required (during burst muscle work, hypoxia, or any energy-depleted state), glycogen phosphorylase activation is initiated by $b \rightarrow a$ conversion. Glycogen synthesis is simultaneously blocked by a direct inhibitory effect of phosphorylase *a* on the production of synthetase *a*. This arises because synthetase phosphatase (which activates synthetase by $b \rightarrow a$ conversion) is inhibited by phosphorylase *a*; it remains inactive unless phosphorylase *a* is removed (experimentally, by specific antibodies, for example) or unless phosphorylase *a* is converted to the *b* form (inactive). At this time, the enzyme can be inhibited again by re-adding phosphorylase *a*. Thus, the domi-

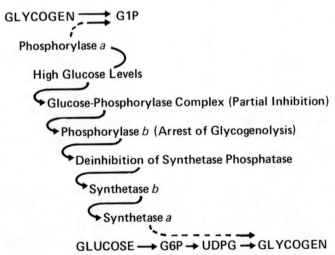

Figure 5.3 Currently postulated sequence of events in liver during glucose-mediated transition from net glycogenolysis to net glycogen synthesis. (Adapted from Hers 1976.)

nant factor controlling glycogen metabolism is the relative concentration of phosphorylase a, not only because it catalyzes the initial rate-limiting step in glycogenolysis but also because it regulates glycogen synthesis by blocking synthetase $b \rightarrow a$ conversion. Nevertheless, once glycogen phosphorylase is in its active form, its activity still is influenced by substrate and modulator concentrations. In muscle, the affinity of phosphorylase a for glycogen is unusually high, some fifty times higher than in the case of the brain enzyme, and the muscle enzyme can therefore bind substrate with high efficiency. Moreover, the enzyme is further activated by adenosine monophosphate (five-fold activation with 0.1 mM AMP) which increases enzyme-substrate affinity still further. These kinetic properties favor glycogen phosphorylase function in the glycolytic direction despite the thermodynamically "uphill" nature of the reaction and can set the stage for a glycolytic activation in muscle which surpasses by far that in any other tissue in the vertebrate body. To achieve this, the regulation of phosphorylase must be at least partially integrated with that of the next key control site in the pathway, 6-phosphofructokinase.

The Control of Phosphofructokinase

Phosphofructokinase (PFK) catalyzes the first committed step in glycolysis and hence has long been recognized as perhaps the single most important control site in glycolysis. A sufficient number of vertebrate tissues and species have now been examined to reveal a common control principle underlying all vertebrate PFKs. Thus for this enzyme the two cosubstrates, fructose 6-phosphate (F6P) and ATP, serve both as substrates per se and as metabolite modulators. One substrate, F6P, behaves as a typical positive modulator, while the cosubstrate, ATP, behaves as an important negative modulator. Fructose 6-phosphate saturation curves are sigmoidal, indicating substrate activation of PFK, and are strongly influenced by modulators, including ATP (table 5.5); ATP saturation curves are hyperbolic, showing substrate inhibition that can be reversed by F6P and other positive modulators.

An indication of what kinetic properties are or are not

Table 5.5 Metabolic effectors of animal phosphofructokinases.

Activators	Inhibitors
Fructose 6-phosphate	ATP
Fructose biphosphate	Citrate
Glucose biphosphate	Phosphoenol pyruvate
AMP	Creatine-P
ADP	Arginine-P (in cephalopods)
Cyclic AMP	3-P-glycerate
K^+	2-P-glycerate
NH_4^+	2,3-diP-glycerate
PO_4^-	NADH (in cephalopods)

adaptable can be obtained by a brief examination of slime mold phosphofructokinase. For cells at late stages of aggregation, when the mold is dependent on amino acids and proteins, glucose is not essential for energy metabolism. Most exogenous glucose in fact is converted to glycogen. Glycolysis (and therefore PFK) does not play a central role in energy metabolism here, but rather serves to supply the cell with trioses and acetate units for synthesis.

Kinetic studies of this kind of PFK indicate that

1. F6P and ATP saturation curves are both hyperbolic with no indication of substrate activation;
2. neither ATP nor citrate are inhibitory;
3. neither AMP nor P_i are stimulatory;
4. FBP and ADP are product inhibitors at the active site.

These are instructive studies for they unequivocally establish that the regulatory (or allosteric) features of phosphofructokinase are not fixed by chemical necessity; they clearly are highly sensitive to selection and adaptation. Just how sensitive is best illustrated by the high degree of regulatory specificity invested in the PFK locus in different tissues and organs.

To illustrate the need for different kinds of phosphofructokinase regulation, let us consider the brain, skeletal muscle, and liver. The brain has a fairly constant demand for glucose that does not vary greatly at any level of mental activity. Glycolysis in skeletal muscle, on the other hand, varies greatly with burst activity, while the liver runs a very modest amount of glucose through this pathway and, in fact, it usually runs the pathway in reverse to synthesize glucose.

Clearly, there is great specificity in the control requirements for PFK. Such specificity is achieved in two ways: through the occurrence of different isozyme forms of PFK differentially sensitive to regulatory metabolites, and through the occurrence of different concentration ranges of the regulatory metabolites in different tissues.

With respect to isozymes, three different types of phosphofructokinases are found in mammalian tissues: isozyme A of heart and skeletal muscle, isozyme B of liver and erythrocytes, and isozyme C present as hybrids with isozyme A in the brain. All three show different responses to known regulatory metabolites of which the adenylates, creatine-P, substrates and products, AMP, P_i, and NH_4^+ are probably the most important (table 5.5).

Adding to isozyme-based regulatory diversity is the occurrence of these metabolites in tissue-specific levels. For example, the level of ATP in muscle is about 5 to 7 mM as compared to about 2 to 3 mM in erythrocytes, liver, and brain. Creatine phosphate is present in skeletal muscle at a concentration of about 30 mM, whereas the level of this metabolite in brain is about 2 mM and none is present in liver and erythrocytes. In contrast, 2,3-diphosphoglycerate exists in appreciable amounts (5 to 10 mM) only in erythrocytes. Obviously the differences in the levels of metabolites among tissues can provide much of the specificity in regulation that is observed. This is readily illustrated by the control of PFK isozyme A in skeletal muscle and heart, the first a relatively anaerobic organ, the second the most aerobic organ in the body, with the possible exception of the brain.

In normal skeletal muscle, after the activation of glycogen phosphorylase and myofibrillar adenosine triphosphatase, rising concentrations of F6P, FBP, adenosine diphosphate (ADP), AMP, and inorganic phosphate coupled with falling concentrations of creatine phosphate and ATP can lead to a very large flare-up of phosphofructokinase when it is required. A central feature of this control system is that two of the positive modulators (FBP and ADP) are products of the reaction; one of the positive modulators (F6P) is, of course, a substrate. Taken together, their regulatory effects lead to an autocatalytic, exponential rate of change from low activity states to high activity states, a characteristic that often is not seen in other cells and that helps to explain the speed with which muscle glycolysis is "turned on."

In the presence of AMP (or other modulators) the $K_{m(F6P)}$/[F6P] is low (because of the low K_m and relatively high F6P levels. Compared to the liver isozyme B, the affinity for F6P is high (about two-to-three-fold higher than in liver). Aside from these characteristics, other distinguishing features of muscle phosphofructokinase compared to those in tissues such as the liver include an overall "tighter" control by most organophosphate modulators and by citrate but a highly reduced sensitivity to ATP inhibition. Thus, for a given percentage change, muscle phosphofructokinase requires about one-tenth as much ADP, AMP, or citrate as does the liver homologue; at the same time, it requires two to three times higher ATP concentrations to bring about the same percentage inhibition.

In heart, the major differences in control arise from lower creatine phosphate levels, lower F6P and FBP levels, and possibly lesser fluctuation in ATP levels. Although the heart and muscle both display the same PFK isozyme form, regulatory specificity thus is effectively achieved by the occurrence of a different microenvironment. This may include metabolites per se or other factors such as pH, inorganic ions, or organic solutes.

Phosphofructokinase Integration with Pyruvate Kinase

A key microenvironmental factor that correlates with skeletal muscle activation is a falling pH due to lactate accumulation. Not surprisingly, muscle phosphofructokinase seems well attuned to modest pH changes; a drop of 0.3 pH units (pH 7.4 to 7.1, for example) leads to dramatic changes in ATP and creatine phosphate sensitivity and to a sharp increase in affinity for both substrates. In lower vertebrates, dropping pH also automatically activates pyruvate kinase, which here typically displays an acidic pH optimum. This effect is potentiated by fructose biphosphate. In lower vertebrates (fishes and reptiles), muscle pyruvate kinase is an allosteric enzyme under close metabolite regulation. Fructose biphosphate, a product of the PFK reaction, serves as a potent feedforward activator of muscle PK, assuring nearly simultaneous activation of both enzymes. In most mammals, in contrast, the major integration mechanism merely involves adenylate coupling; that is, ADP, the product of the phosphofructokinase reaction, is a substrate for pyruvate kinase, and this in itself serves to automatically coordinate the activities of these two enzymes. With the activation of PK, one final control function (that of redox balance) is necessary for completing the process of anaerobic glycolysis.

Maintenance of Redox Balance during Anaerobic Glycolysis

In general, the redox balance (the ratio of oxidized to reduced nicotinamide adenine dinucleotide, $NAD^+/NADH$) during anaerobic glycolysis is maintained by a functionally 1:1 activity ratio between glyceraldehyde 3-phosphate dehydrogenase and lactate dehydrogenase, and this surely is the situation during normal rates of glycolysis. Both enzymes occur in higher titer in vertebrate muscle than in other tissues, and in rat muscle, for example, they occur in about a 1:1 ratio. In extreme situations, the lactate dehydrogenase reaction becomes limiting by mass action effects; at this time, pyruvate and lactate accumulate in constant proportion (lactate to pyruvate ratio of about 20), and if muscle glycolysis is to continue, it must have some other source of NAD^+. That source is usually considered to be the α-glycerophosphate dehydrogenase reaction, in which case, α-glycerophosphate would accumulate as an additional anaerobic end product. The degree to which this occurs and the way in which α-glycerophosphate dehydrogenase function is integrated with the function of lactate dehydrogenase is only now being clarified. Current impressions of the field, however, imply relatively exclusive functions of these two enzymes (see chapter 6), so there may be other, thus far

78 LIVING WITHOUT OXYGEN

unidentified, mechanisms for sustaining glycolysis when the LDH reaction becomes limiting by mass action effects.

Suggested Readings

ATKINSON, D. E. 1977. *Cellular Energy Metabolism and Its Regulation.* New York: Academic Press.

BORGMANN, U., LAIDLER, K., and MOON, T. W. 1975. Kinetics and thermodynamics of lactate dehydrogenases from beef heart, beef muscle, and flounder muscle. *Can. J. Biochem.* 53:1196–1206.

BORGMANN, U., and MOON, T. W. 1975. A comparison of lactate dehydrogenase from an ectothermic and an endothermic animal. *Can. J. Biochem.* 53:998–1004.

BAUMANN, P., and WRIGHT, B. E. 1968. The phosphofructokinase of *Dictyostelium discoideum. Biochemistry* 7:3653–3661.

EDINGTON, D. W., WARD, G. R., and SAVILLE, W. A. 1973. Energy metabolism of working muscle: concentration profile of selected metabolites. *Am. J. Physiol.* 224:1375–1380.

EVERSE, J., and KAPLAN, N. O. 1973. Lactate dehydrogenases: structure and function. In Meister, A., ed., *Advances in Enzymology*, vol. 37. New York: Wiley. Pp. 61–133.

FERSHT, A. 1977. Enzyme structure and mechanism. Reading: Freeman.

HERS, H. G. 1976. The control of glycogen metabolism in the liver. *Annu. Rev. Biochem.* 45:167–190.

HOCHACHKA, P. W., FRENCH, C. J., and BALLANTYNE, J. S. 1979. Thermal modulation of binding and catalytic functions of enzymes. In Bhatnagar, R. S., ed., *Molecular Basis of Environmental Toxicity.* Ann Arbor: Ann Arbor Science Publishers. Pp. 505–513.

HOCHACHKA, P. W., and STOREY, K. B. 1975. Metabolic consequences of diving in animals and man. *Science* 187:613–621.

MIWA, S., NAKASHIMA, K., and SHINOHARA, K. 1975. Physiological and pathological significance of human pyruvate kinase isozymes in normal and inerited variants with hemolytic anemia. In Markert, C. L., ed., *Isozymes*, vol. 2, *Physiological Function.* New York: Academic Press. Pp. 487–500.

PENHOET, E. E., KOCHMAN, M., and RUTTER, W. J. 1969. Molecular and catalytic properties of aldolase C. *Biochemistry* 8:4396–4402.

TSAI, M. Y., and GONZALEZ, F., and KEMP, R. G. 1975. Physiological significance of phosphofructokinase isozymes. In Markert, C. L., ed., *Isozymes.*, vol. 2, *Physiological Function.* New York: Academic Press. Pp. 819–835.

WILLIAMSON, D. H., and BROSNAN, J. T. 1974. Concentration of metabolites in animal tissues. In Bergmeyer, H. D., ed., *Methods of Enzymatic Analysis*, vol. 4. New York: Academic Press. Pp. 2266–2302.

Chapter Six

Integrating Aerobic and Anaerobic Glycolysis

Many tissues express significant capacities for both anaerobic and aerobic glycolysis, and a problem which has remained latent to this point in the discussion is how the two processes are integrated. One important reason a problem arises is that mitochondria are impermeable to NADH. Because of this impermeability, in all cells that generate NADH in the cytosol during aerobic glycolysis, there is a requirement for hydrogen shuttles to move reducing equivalents to the electron transfer system (ETS). In mammalian tissues, two such shuttles seem the most likely candidates for this role. The major one is the malate–aspartate shuttle in which the component enzymes are cytosolic and mitochondrial forms of both aspartate aminotransferase and malate dehydrogenase, and at least two exchange mechanisms (malate in for 2-ketoglutarate out; glutamate in for aspartate out):

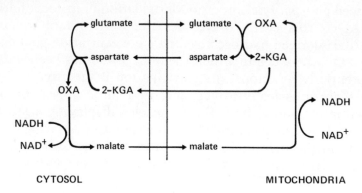

According to current evidence, the transport mechanisms for malate, 2-ketoglutarate, and glutamate are potentially bidirectional and occur with no significant energy cost. Aspartate transport, on the other hand, appears to be an energy-dependent system; while reversible under some conditions, it is considered to be largley unidirectional, favoring aspartate efflux from the mitochondria. These features of the aspartate carrier mechanism may be necessary to allow the movement of reducing equivalents "uphill," since the NADH/NAD$^+$ ratio in the mitochondria is known to be substantially more reduced than in the cytosol.

A simpler mechanism for transferring NADH-derived hydrogen into the mitochondria is the α-glycerophosphate (α-GP) cycle. In this cycle, DHAP is reduced by cytosolic NADH to α-GP, a reaction catalyzed by cytosolic α-GPDH; the α-GP is oxidized by a second enzyme, α-GP oxidase, located on the inner mitochondrial membrane:

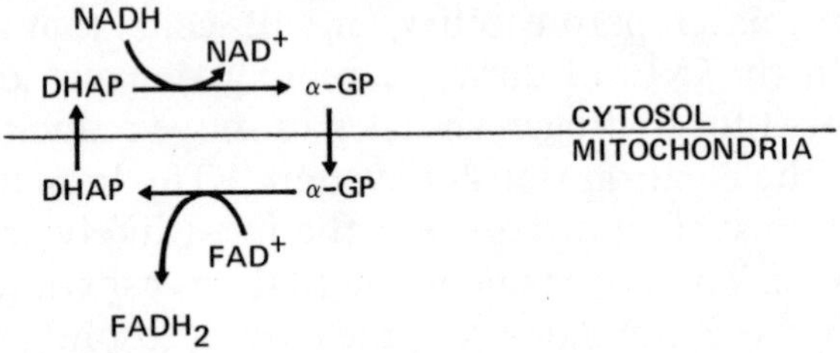

The FAD-linked α-GP oxidase donates electrons to the electron transfer system at the level of cytochrome b. Hydrogen transfer by this shuttle, therefore, is associated with a reduced phosphorylation ratio for oxidation of mitochondrial NADH; this easily overcomes the difficulty of moving NADH-derived hydrogen against a concentration gradient.

Of all tissues in which the α-GP cycle has been found potentially functional, its participation is most prominent in flight muscles of insects and the mantle muscle of fast swimming squid, both these muscles displaying a strong dependence upon carbohydrate as a source of carbon and energy. In the mammalian heart, which burns carbohydrate and fat, the malate–aspartate cycle is the more active shuttle mechanism, the α-GP cycle playing only a minor role even when the α-GP oxidase activity is increased by thyroid administration. However, the α-GP cycle is demonstrated in rat heart when α-GP levels and cytosolic NADH/NAD$^+$ ratios are high, conditions that may be expected during transition from hypoxic to normal aerobic metabolism.

Three other hydrogen shuttles are known. A fatty acid shuttle can be reconstructed with isolated liver mitochondria and is consistent with the following scheme:

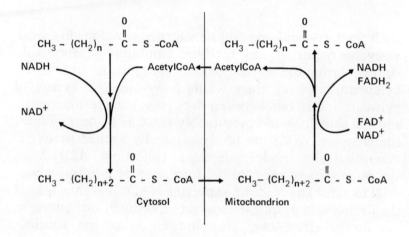

The proline shuttle can be reconstructed with mitochondria isolated from blow fly flight muscle:

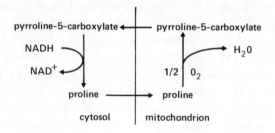

A pyruvate–lactate shuttle can be theoretically set up in mammalian and avian sperm since sperm-specific LDH-X occurs intramitochondrially:

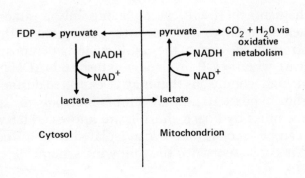

Both the fatty acid and proline shuttles could readily work as cycles requiring only catalytic amounts of carbon substrates; during the aerobic oxidation of glucose or fructose by sperm, however, there would be no need for cycling of pyruvate and lactate between the cytosol and mitochondria and the shuttle would presumably serve as a linear path for the complete oxidation of glycolytically formed pyruvate. Parenthetically, under anaerobic conditions, LDH-X in sperm appears to serve in the formation of intramitochondrial lactate. The reaction generates NAD^+ and thus allows the Krebs cycle to spin at a low rate and the mitochondria to sustain substrate-level phosphorylation at the succinic thiokinase step.

Although the fatty acid, proline, and lactate-based shuttles can be reconstructed in isolated mitochondria, it is uncertain under what physiological conditions they operate. Or, in fact, whether they operate as shuttles at all under any physiological conditions. In many tissues, the enzymic components of the malate-aspartate and α-GP cycles occur at much higher levels than the enzymes of the fatty acid or proline shuttles, while intramitochondrial LDH is found commonly only in sperm. For this if for no other reason, it is probable that the malate–aspartate and the α-GP cycles are the commonest mechanisms for transferring reducing equivalents across the mitochondrial boundary. No matter which of these two hydrogen shuttles is utilized, in tissues displaying both anaerobic and aerobic glycolysis, it is critically important that the flow of hydrogen through cytosolic dehydrogenase reactions be closely regulated. In particular, malate dehydrogenase (MDH) and α-glycerophosphate dehydrogenase (α-GPDH) function should be curtailed during anaerobic glycolysis, while lactate dehydrogenase (LDH) function should be curtailed during aerobic glycolysis. How this is achieved has not been fully clarified. However, some insight into this overlooked aspect of metabolic regulation can be gleaned from kinetic studies of lactate and glycerophosphate dehydrogenases.

Properties of α-GPDH in Muscles Lacking LDH

To get at the question of how α-GPDH is regulated, it is advantageous to analyze the catalytic properties of enzyme homologues in tissues where the sole function of cytosolic α-GPDH is to shuttle hydrogen and where LDH is either totally or almost totally deleted. This situation is well established in bee flight muscle and also seems to prevail in the mantle muscle of pelagic squid. A kinetic analysis indicates that in both the bee and the squid, α-GPDH has a low affinity for DHAP but an extremely high NADH affinity (table 6.1). The enzyme is thus adjusted to spark the α-GP

Table 6.1 Kinetic properties of "unidirectional" α-glycerophosphate dehydrogenases from the squid and bee compared to those from vertebrate white muscles.

Species/tissue	Kinetic constant	Kinetic constant (mM) for			
		DHAP	NADH	NAD$^+$	α-GP
Bee flight muscle	K_m	0.24	0.005	1.6	2.0
	K_i	0.005	0.001		10.0
Squid mantle	K_m	0.15	0.015	0.14	1.0
	K_i	0.012	0.004		over 15
Tuna white muscle	K_m	0.072	0.014	0.4	0.25
	K_i	0.068	0.007	1.1	0.25
Seal muscle	K_m	0.14	0.015	0.6	0.5
	K_i	0.08	0.005	2.8	1.2
Turtle white muscle	K_m	0.09	0.015	0.25	0.3
	K_i	0.09	0.003	0.5	1.0

Source: Hochachka and Guppy (1977).

cycle whenever DHAP becomes available. Since the NADH/NAD$^+$ pool size is only about 1 mM, the problem of NADH levels rising without limit when and if the enzyme becomes saturated does not arise even if the enzyme displays a very low K_m for NADH. The DHAP pool size, on the other hand, is less limited, and a high K_m for DHAP assures that high rates of cycling occur only when glycolytic rates (and hence DHAP levels) are elevated. A low DHAP-affinity also prevents the enzyme from saturating with DHAP, which means that the more DHAP formed, the higher the α-GP cycling rates attainable, a feature of obvious advantage in highly active tissues such as squid mantle and bee flight muscles.

Equally important, the product, α-GP, is a poor inhibitor of α-GPDHs in the bee and squid muscles. That is why α-GP accumulates in these muscles when they are made anoxic. As we shall show, this cannot occur with the vertebrate enzyme because of potent α-GP inhibition. However, an important consequence of an α-GP-*insensitive* α-GPDH is that during burst muscle work α-GP levels may be expected to increase to a higher steady-state level, allowing α-GP oxidase to work at closer to its maximum velocity; alternatively, a sudden burst of cytosolic α-GPDH might lead to a pulse of α-GP production so as to activate α-GP oxidase. Either way, α-GP-insensitive α-GPDHs seem advantageous in these systems.

A low affinity for α-GP contributes to another key element of the bee and squid enzymes: unidirectional catalysis in vivo. This depends in part on properties of the back reaction. When the back reaction is considered, both enzymes are very sensitive to product inhibition, particularly with respect to NADH. In both cases, NADH is a competitive inhibitor with respect to NAD$^+$ (the K_i values are 1.0 and 4.0 μM for the bee and squid enzymes, respectively). Similarly, DHAP is a competitive inhibitor with respect to α-GP, and the K_i values for DHAP are also low, about $\frac{1}{10}$ the respective K_m values for the forward reaction (table 6.1). Operationally, what this means is that in both systems, when the forward reaction is half-saturated with either DHAP or

NADH, the back reaction is strongly inhibited by either or both metabolites. Therefore, whenever one or more components occur at limiting concentrations, one-way in vivo function, DHAP $\rightarrow$ α-GP, seems most probable. This conclusion is bolstered by thermodynamic considerations (for the reaction proceeds in the forward direction with a large, negative free-energy change), and by the respective pH optima for the forward and backward reaction (since the back reaction is reduced two-to-four-fold at physiological pH compared to optimal pH).

In summary, then, four kinetic properties appear to characterize α-GPDHs exclusively specialized for hydrogen shuttling: a high NADH affinity, a low DHAP affinity, an insensitivity to α-GP product inhibition, and essentially unidirectional catalysis at low substrate concentrations. Are these features retained or modified in α-GPDH homologues in tissues with both aerobic and anaerobic capacities? To facilitate exploration of this question it is convenient to work with a tissue in which both oxidative and glycolytic capacities are exaggerated. One such tissue is the white myotomal muscle of tuna.

Tuna White Muscle, a Model System

Tuna are, in many ways, "ultimate teleosts." They maintain their musculature at 29 to 35°C, which is 3 to 10°C higher than those parts of the body which are at ambient temperature. They can sustain for an indefinite period a speed of 3 to 5 lengths/second, and can swim at up to 20 lengths/second for short periods. These bursts, which represent absolute velocities up to 40 mph, can be maintained for about 10 minutes with 30 second periods between bursts. The outstanding swimming speeds of tuna are powered by a musculature divided clearly into areas of white and red fibers. Even if the relative mass and position of red and white muscles in tunas are somewhat unique, the metabolic design of the myotome seems typical, in that white muscle is inactive at low swimming speeds but is increasingly recruited at higher tail beat frequencies. Enzyme and ultrastructure studies show the red muscle to have more mitochondria than white muscle and therefore higher specific activities of the Krebs cycle enzymes; for glycolytic enzymes, the reverse pattern is observed (table 6.2).

Whereas tuna white muscle may be qualitatively quite ordinary, quantitatively it is extraordinary. This muscle generates the outstanding anaerobic burst described above and not surprisingly displays the highest LDH activity thus far found in nature, nearly 6000 units of LDH per g at 25°C. As a result, the muscle accumulates lactate to nearly 100 μmol/g wet wt during such (10-minute) burst exertions. It is

Table 6.2 Enzyme levels in skipjack tuna white muscle, red muscle, and heart, expressed in terms of μmols substrate converted/gm tissue/min at 25°C. Average values from four tuna.

Enzyme	Red muscle	White muscle	Heart
Glycogen phosphorylase	22.0	106.2	9.6
Phosphoglucomutase	31.3	152.8	13.3
Phosphoglucoseisomerase	84.4	426.0	115.9
Hexokinase	1.0[a]	0.8	4.7
Phosphofructokinase	10.0[a]	25.0	20.6
Aldolase	35.5	269.2	22.5
Triose phosphate isomerase	1414.6	11413.0	1661.1
α-Glycerophosphate dehydrogenase	21.7	104.5	6.1
Phosphoglycerate kinase	371.1	1982.7	301.4
Enolase	77.7	522.4	42.6
Pyruvate kinase	195.2	1294.9	126.6
Lactate dehydrogenase	514.4	5492.3	449.0
Citrate synthase	20.6	2.7	25.8
Malate dehydrogenase	723.4	718.0	884.2
Glutamate dehydrogenase	5.9	3.0	7.7
Glutamate-oxaloacetate transaminase	101.9	43.0	155.0
Glutamate-pyruvate transaminase	7.7	2.0	13.1
Creatine phosphokinase	554.2	516.4	115.2
Myokinase	381.8	946.1	240.2

Source: Data from Hochachka, Hulbert, and Guppy (1978).

[a] The hexokinase and phosphofructokinase values are unreliable due to high blank activities and extreme instability.

instructive that such a powerful anaerobic system occurs in a tissue simultaneously displaying a significant aerobic potential, primed by an α-glycerophosphate cycle. Cytoplasmic α-GPDH occurs in unusually high activities for muscle and is thought to function in supplying NAD^+ for aerobic glycolysis. Tuna white muscle thus accentuates dual metabolic function (aerobic and anaerobic glycolysis) but also exposes two serious problems with such a design. One is expected; the other is not. In the first place, if α-GPDH is not "turned off" during anaerobiosis, a deleterious drain of carbon from mainline glycolysis may result. This is the expected problem. The unexpected problem is how to control LDH activity. If the activity of LDH is not curbed during aerobic work, there could be serious competition for NADH and/or a drain of pyruvate into lactate.

Controlling Tuna Muscle α-GPDH

In view of the patent need to prevent simultaneous function of α-GPDH and LDH, it is particularly interesting that in tuna white muscle, α-GPDH shows many of the same kinetic properties that we already have seen in the bee and squid enzymes. In both kinds of systems, the activities are high (about 100 units in tuna, compared to 200 to 250 units in the bee and squid muscles); the forward reaction shows a sharp pH optimum at pH 7 while the reverse reaction shows an alkaline pH optimum; the K_m for DHAP (0.07 mM) is relatively high while the K_m for NADH (0.014 mM) is low. Most of the features favoring unidirectional catalysis are also at least partially preserved. Hence on all these counts tuna muscle α-GPDH seems similar to its homologues in bees and squids, and seems fully capable of functioning within the α-GP cycle. When the effects of α-GP on the enzyme are considered, however, the similarities break down.

The constraints on the type of α-GPDH in white muscle do not appear until we consider how α-GPDH is turned off. In this connection, despite diligent search for potential regulatory metabolites and for mechanisms by which α-GPDH activity can be blocked, in the bee and the squid ATP appears as the only possible candidate. K_i values are 2.0

and 0.6 mM for the bee and squid enzymes, respectively, values well within the physiological range. Through this means, α-GPDH activity can be integrated with the overall energy status of the cell. In this respect, the enzyme in tuna white muscle is again similar, the K_i for ATP being about 1.5 mM.

However, in tissues such as tuna white muscle, adenylate regulation by and of itself is clearly inadequate, for here the fundamental kinetic and metabolic requirement is to turn α-GPDH off when LDH is on (that is, when the system is anaerobic, as during burst swimming). That requirement cannot be supplied by ATP and is absent in muscles with little or no anaerobic capacity; however, it can be met by any metabolite accumulating during anaerobic conditions. That is why importance is attached to the observation that tuna white muscle α-GPDH is extremely sensitive to α-GP, since a good (perhaps the best) metabolite signal for turning off α-GPDH when O_2 is limiting is α-GP accumulation. That also is why α-GPDHs in all skeletal muscles thus far examined (rat, rabbit, tuna, porpoise, seal, turtle) display a relatively high sensitivity to α-GP (table 6.3). If, as is believed, the function of α-GP inhibition is to turn off α-GPDH whenever the tissue goes anoxic (thus preventing a carbon and energy drain on glycolysis), the development of

Table 6.3 Kinetic constants of "reversible" α-glycerophosphate dehydrogenases from rabbit liver compared to the enzyme in tuna red muscle and muscle from the dolphin, *Delphinus delphis*.

Species/tissue	Kinetic constant	Kinetic constant (mM) for			
		DHAP	NADH	NAD$^+$	α-GP
Rabbit liver	K_m	0.02	0.004	0.2	0.68
	K_i	0.02	0.004	0.15	0.65
Tuna red muscle	K_m	0.05	0.007	0.15	0.2
	K_i	0.10	0.002	0.8	0.75
Dolphin muscle	K_m	0.02	0.005	0.17	0.15
(*Delphinus*)	K_i	0.05	0.002	1.2	0.7

Source: Hochachka and Guppy (1977).

this feature is perhaps a predictable design. But what of the reverse of this problem? How is LDH function controlled when α-GPDH function is required in aerobic glycolysis? Again, the question and its answer are best illustrated in studies of tuna white muscle, where they were first recognized.

Controlling Tuna Muscle LDH

From currently available data at least three factors appear to influence LDH activity in tuna white muscle: pH, creatine phosphate, and temperature. A drop in pH leads to a drop in the K_m for pyruvate, a feature that may be potentiated by elevated temperatures which promote low pH; dropping creatine phosphate levels deinhibit the enzyme; and a rise in temperature leads to increased affinity for NADH, plus, of course, the usual Q_{10} effect. Taken together, the kinetic date allow the construction of a plausible *modus operandi* of how LDH activity can be confined to situations which do not favor α-GPDH activity. Under "resting" aerobic conditions, the pH and creatine phosphate levels are relatively high (over pH 7 and over 20 mM, respectively), while temperature is relatively low; α-GPDH activity is favored while LDH is dampened because of (1) differing pH optima, (2) a low LDH affinity for pyruvate, and (3) creatine phosphate inhibition of LDH. During burst swimming, which is powered primarily by an intense anaerobic glycolysis, creatine phosphate levels drop dramatically (to about 1 mM), pH drops, and temperature rises somewhat (by about 5°C). LDH activity is now strongly favored because of creatine phosphate deinhibition and a pH-induced increase in LDH affinity for pyruvate. In contrast, α-GPDH activity is dampened by α-GP product inhibition and low pH.

During steady-state swimming at sea, which is powered at least in part by an aerobic glycolysis, pH, creatine phosphate levels, and temperature are all relatively high. Under these conditions, LDH activity is dampened by high pH and creatine phosphate inhibition. Moreover, the 10°C excess muscle temperature leads to a fall in LDH affinity for pyruvate which, unlike the situation in burst swimming, is not matched by a concomitant rise in pyruvate levels.

Thus, from strictly kinetic studies, it is possible to visualize a relative exclusive function of either LDH or α-GPDH, allowing a tissue obviously geared for an impressive anaerobic glycolysis to sustain low but significant aerobic glycolysis during aerobic swimming (fig. 6.1). But do these mechanisms actually work in vivo? The answer is yes, and the evidence comes from two kinds of studies: end product accumulation and two-enzyme competition experiments.

Lactate and α-GP Accumulation

That mutually exclusive function of LDH and α-GPDH is evidently widely achieved is most convincingly demon-

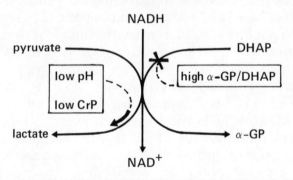

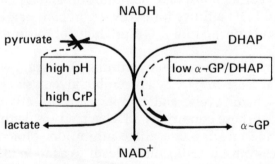

Figure 6.1 Competition between lactate and α-glycerophosphate dehydrogenases.

strated by metabolite measurements which show that in hypoxic or anoxic stress lactate always accumulates to much higher levels than does α-glycerophosphate, even though both LDH and α-GPDH are present in the cell. In the ischemic perfused rat heart, for example, lactate rises to about 20 μM/gm, while α-GP increases to only about 2 μM/gm. Lactate and α-GP increase to about 30 and 3 μM/gm, respectively, during anaerobic work in isolated rat skeletal muscle. Lactate and α-GP in tuna white muscle increase to 100 and 4 μM/gm, respectively, during 10-minute bursts of swimming. In some of these cases, especially in tuna white muscle, the level of αGPDH is easily high enough to channel a large fraction (up to $\frac{1}{2}$) of glucose-derived carbon and hydrogen to α-GP. Yet this does not occur. Similarly, during aerobic glycolysis, when the α-GP cycle is operative, the activity of LDH is potentially far in excess of any other step in the overall metabolism of glucose, and in theory lactate could accumulate. Yet this too does not occur. The kinetic features noted above are the reason why these things are so, but proof that the metabolite changes are large enough requires direct tests of competitive abilities of LDH and α-GPDH. In such tests, advantage is taken of varying sensitivities of LDHs to creatine phosphate, and varying sensitivities of different α-GPDHs to α-GP.

Lactate Dehydrogenase versus
Alpha-glycerophosphate Dehydrogenase

The creatine phosphate sensitivity of different LDHs appears to roughly correlate with the oxidative capacity of the tissue from which they are derived. The sources of LDH listed in table 6.4, for example, are arranged approximately in order of increasing oxidative capacity. This correlation is probably secondary and derives from the fact that the LDH isozyme function and content vary in these tissues. The creatine phosphate sensitivity appears to depend upon the relative abundance of LDH subunits displaying M-type versus H-type properties. Pure M_4 LDH is one of the least creatine-phosphate-sensitive preparations studied while pure H_4 LDH is one of the most sensitive. Therefore, in the presence of creatine phosphate, different LDHs should show differing capacities to compete with α-GPDH for a common source of NADH.

That this indeed is the case is evident in two-enzyme competition experiments between rabbit muscle α-GPDH and either of two (H_4 and M_4) kinds of LDH enzymes in the presence and absence of creatine phosphate. Similar initial activities of both dehydrogenases lead to similar contributions to total NADH oxidation in both cases in the absence of creatine phosphate. As expected, creatine phosphate has little or no effect on the fraction of NADH oxidized by M_4 LDH, a creatine-phosphate-resistant isozyme. In contrast, about three times more NADH is oxidized by α-GPDH compared to H_4 LDH when 20 mM creatine phosphate is included. Thus under conditions of limiting NADH, creatine phosphate at physiological concentrations is an important modulator of LDH contribution to redox regulation. But is α-GP accordingly effective in α-GPDH regulation?

Table 6.4 Inhibition of lactate dehydrogenase by 20 mM creatine phosphate. Assay conditions: 0.1 mM pyruvate, 0.1 mM NADH, 25°C, pH 7.0.

Source of LDH	Type of LDH function	% of inhibition by 20 mM creatine phosphate
Hoplias white muscle[a]	Pyruvate reductase	25
Turtle white muscle	Pyruvate reductase	26
M_4 from rabbit muscle	Pyruvate reductase	29
Hoplerythrinus white muscle[a]	Pyruvate reductase	32
Skipjack white muscle	Pyruvate reductase	38
Hoplias heart[a]	Bifunctional	39
Arapaima heart[a]	Bifunctional	39
Hoplerythrinus heart[a]	Bifunctional	41
Osteoglossum heart[a]	Bifunctional	47
H_4 from beef heart	Lactate oxidase	71
Rat brain	Lactate oxidase	71
Weddell seal heart	Lactate oxidase	77

Source: Guppy and Hochachka (1978b).

[a] 0.3 mM pyruvate in the case of the Amazon fishes.

Table 6.5 Competition for NADH between rabbit muscle α-glycerophosphate dehydrogenase and different lactate dehydrogenase isozymes.

| | 0.0 mM creatine phosphate | | 20 mM creatine phosphate | |
LDH isozyme	% of total oxidation by LDH	% of total oxidation by α-GPDH	% of total oxidation by LDH	% of total oxidation by α-GPDH
H$_4$	52.4	47.7	25.6	74.3
M$_4$	49.5	50.6	48.3	51.7

Source: Guppy and Hochachka (1978b).

To answer this question, advantage is taken of α-GP sensitive and α-GP insensitive forms of α-GPDHs obtained from different species. As already indicated, the kinetic data suggest that α-GP resistant α-GPDHs should be more competitive with LDH under conditions of high α-GP levels than would be the typical vertebrate, α-GP-sensitive homologue. A direct test of this hypothesis is available through two-enzyme competition experiments using one LDH form (H$_4$) and two types of α-GPDHs (table 6.5). Rabbit muscle α-GPDH displays a high sensitivity to α-glycerophosphate inhibition while the honey bee enzyme is resistant to product inhibition (table 6.1). In the absence of α-GP, the two forms of αGPDH compete with similar effectiveness for NADH. The two contributions to total NADH oxidation are not exactly equal (about 42% versus 50% of total NADH oxidation by the rabbit and bee α-GPDHs, respectively), possibly because of differences in K_m values for substrates and NADH. The similarities in behavior of the two enzymes disappear in the presence of 2 mM α-glycerophosphate. Under these conditions, LDH contribution to NADH oxidation exceeds rabbit muscle α-GPDH contribution by nearly ten-fold, but it exceeds the oxidation due to bee muscle α-GPDH only marginally (table 6.6). It is worth recalling that during anaerobic work in mammalian muscle, α-GP levels rise to about 3 μM/gm, somewhat higher than the concentrations used in the above competition experiments. Thus, with both LDH and α-GPDH competing for the same limit-

Table 6.6 Competition for NADH between H_4 lactate dehydrogenase and different α-glycerophosphate dehydrogenase isozymes.

α-GPDH source	0.0 mM α-glycerophosphate		2.0 mM α-glycerophosphate	
	% of total oxidation by LDH	% of total oxidation by α-GPDH	% of total oxidation by LDH	% of total oxidation by α-GPDH
Rabbit muscle	58.6	41.5	89.0	12.6
Honey bee flight muscle	50.0	50.9	65.7	36.4

Source: Guppy and Hochachka (1978b).

ing pool of NADH, α-GPDH sensitivity to reaction product (α-GP) greatly reduces the wasteful channeling of carbon and hydrogen from "mainline" glycolysis into α-GP.

In Vivo NADH Concentrations

If these interpretations are correct, NADH, at least under some conditions, must occur at limiting concentrations. Until recently no reliable estimates of cytosolic NADH were available. Those that were reported ranged between 0.03 and 0.15 μmol/gm wet weight. Such values are for the whole cell and the NADH concentration in the cytosol can only be less. Concentration ratios, $NAD^+/NADH$, in the cytosol of liver, brain, and fibroblasts vary from 7 to 2000. Assuming about a 1 mM pool size, NADH levels can be estimated at between 0.5 and 10 μM. Most of these earlier estimates appear to be too high. Studies using the technique of turbulent flow for the rapid lysis of isolated hepatocytes indicate that the concentration of free NADH in the cytosol is in the 0.06 to 1.5 μM range under differing metabolic conditions (starved versus fed nutritional states, with and without exogenous ammonia). Probably because the NADH binding site of dehydrogenases is conservative, the affinity constants of many dehydrogenases for NADH are remarkably uniform, in the 10 to 20 μM range. A comparison of the two data sets indicates that the affinity constants are higher than the lower limits of NADH concentration in vivo. It is therefore probable that NADH would often, if not always, be limiting in the cytoplasm. At such times, creatine phosphate and α-glycerophosphate regulation of LDH and α-GPDH, respectively, would profoundly influence back and forth transitions from anaerobic to aerobic glycolysis. Whereas this may adequately explain the relatively exclusive function of LDH and α-GPDH, we are still faced with the problem of controlling MDH. Do similar mechanisms integrate MDH and LDH functions during aerobic–anaerobic transitions? Unfortunately, the situation here is not as clear and must vary greatly between highly oxidative and highly glycolytic tissues because MDH occurs in distinct cytosolic and mitochondrial forms. This latter point can be nicely made by expanding our analysis of the tuna model system, in this case contrasting tuna white and red muscles.

MDH Control in Oxidative Tissues: Tuna Red Muscle

At the outset it should be emphasized that in highly oxidative tissues, such as tuna red muscle, MDH may occur at levels that are higher than any other enzyme in energy metabolism except for triose phosphate isomerase (table 6.2). In tuna red muscle, the total MDH activity is about 700 units/gm, which means from 300 to 400 units of MDH activity in the cytosol. It is a fair assumption that under aerobic conditions, cytosolic MDH functions in the malate–aspartate cycle. When O_2 supplies are limiting, however, the electron transfer system slows down, the malate–aspartate cycle appears to become unspanned, and a net flow of carbon is redirected to malate and ultimately succinate, which accumulates in red muscle of fishes under these conditions. The ultimate source of malate, and thus succinate, is thought to be aspartate, whose mobilization is initiated by aspartate aminotransferase. The latter is thought to couple with alanine aminotransferase through 2-KGA and glutamate (fig. 6.2). An identical metabolic pattern occurs in mammalian heart muscle during hypoxia, as best demonstrated by Taegtmeyer.

In such situations, in fish red muscle and in the mammalian heart, the flow of carbon to succinate is presumably

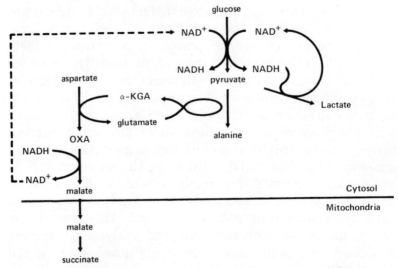

Figure 6.2 Proposed redox coupling between aspartate mobilization to succinate and glucose mobilization to alanine during anoxic stress in oxidative tissues (such as red muscle and heart).

limited by aspartate pools and by pyruvate availability for alanine aminotransferase. However, any loss of glycolytically formed pyruvate to alanine leads to a redox imbalance in anaerobic glycolysis; for the process to continue, NADH must be oxidized by a dehydrogenase other than LDH. That may be the function of MDH, in which event its activity would have to keep pace with alanine, not lactate, formation. But how this is achieved is unknown.

MDH Control in Glycolytic Tissues

The situation is very different in highly glycolytic tissues such as tuna white muscle. Here, total MDH activity is remarkably similar to that in red muscle (table 6.2), but most of it (over 90%) is in the cytosol. Of this large activity, only a small fraction (less than 10%) would be necessary for function in the malate–aspartate cycle. Why then does tuna white muscle retain the rest?

At the moment, the answer to this question is not known. The situation in tuna white muscle, however, is not unique. In white muscle of hypoxia-adapted Amazon fish, MDH is found to occur in very high activities. It is known to occur in high activities in extremely anaerobic-type invertebrate muscles; for example, in oyster adductor and *Nautilus* spadix muscle, MDH occurs at higher activities than any other enzyme in the energy metabolism of these tissues. A similar pattern emerges in enzyme profiles of parasitic helminths. So the condition in tuna white muscle clearly is not unusual, but because of good background information, the tuna system leads us to a testable model of MDH function that may be generally applicable.

An important clue to this problem comes from LDH studies showing that the K_m for pyruvate for tuna white muscle LDH can be very high; at pH 7.4 and biological temperature, it is about 1.3 mM. As anaerobic glycolysis proceeds, conditions progress, allowing an activation of LDH as described above. But how is anaerobic glycolysis to proceed initially without some continued source of NAD^+? How, in other words, is enough pyruvate to be made to allow initiation of the high LDH capacity available to the tissues? One possiblity is that the high MDH activity of tuna white muscle serves this function, that during initial stages of glycolytic activation, cytosolic redox is maintained by aspartate-primed MDH function. This would allow pyruvate levels to accumulate and gradually "turn on" LDH. Once conditions arose favoring LDH function, further MDH function could be curtailed:

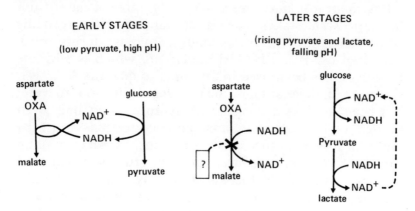

According to this scheme, which is similar in all respects to that in cephalopod muscle (see chapter 4), the catalytic properties of MDH should be such as to favor function during early stages, but not later stages, of glycolysis. At the moment, it is not known how MDH function is "turned off" during later stages of anaerobiosis, when LDH is active. Nevertheless, if the above model of tuna white muscle MDH function is correct, it leads to quite specific and testable predictions concerning anaerobic metabolism in these and comparable tissues. The principal predictions are that during early stages of anaerobic glycolysis, malate is the primary depository of glucose-derived hydrogen. Alpha-glycerophosphate may also accumulate some hydrogen, but it presumably quickly limits its own further formation by α-GPDH inhibition. In later stages of anaerobic glycolysis, lactate is quantitatively the only important depository of glucose-derived hydrogen.

A technique is now available that may allow the testing of this hypothesis directly in vivo. It depends upon the use of glucose specifically labeled with tritium, and is a modification of that used under aerobic conditions to study hydrogen shuttling in the heart and adipose tissue. When $4\text{-}^3\text{H}$-glucose is used, the hydrogen introduced into the aldehyde group of glyceraldehyde 3-phosphate is replaced by tritium; reduction of NAD^+ by GAPDH produces tritiated NADH. Reoxidation of the tritium-labeled NADH by LDH, MDH, or α-GPDH transfers the tritium to malate, lactate, and α-GP without loss into water. Normally, most of the tritium lost into water takes place after transfer of the reducing equivalents into the mitochondria (that is, during mitochondrial α-GP and NADH oxidation), which is why the technique has been used in attempts to quantify hydrogen shuttling. In these studies, the transfer of tritium into water has been used as a measure of hydrogen shuttling, even though under some circumstances corrections must be made for loss of ^3H to water by other means. An error of up to about 20% is introduced in such studies in perfused mammalian heart. In the case of highly anaerobic tissues such as tuna white muscle, this problem can be avoided by looking for NADH-derived ^3H not in water but in the potential end products of anaerobic metabolism, malate, α-GP, and lactate. When that is done it will supply a direct in vivo test of the models described above for the controlled (if competitive) anaerobic functions of cytoplasmic dehydrogenases.

Suggested Readings

BALBONI, E. 1978. A proline shuttle in insect flight muscle. *Biochem. Biophys. Res. Commun.* 85:1090–1096.

CEDERBAUM, A. I., LIEBER, C. S., BEATTIE, D. S., and RUBIN, E. 1973. Characterization of shuttle mechanisms for the transport of reducing equivalents into mitochondria. *Arch. Biochem. Biophys.* 158:763–781.

GUPPY, M., and HOCHACHKA, P. W. 1978a. Controlling the highest lactate dehydrogenase activity known in nature. *Am. J. Physiol.* 234:R136–R140.

————. 1978b. Role of dehydrogenase competition in metabolic regulation: the case of lactate and α-glycerophosphate dehydrogenases. *J. Biol. Chem.* 253: 8465–8469.

GUPPY, M., HULBERT, W. C., and HOCHACHKA, P. W. 1979. Metabolic sources of heat and power in tuna: enzyme and metabolite profiles. *J. Exp. Biol.*, 82:303–320.

HOCHACHKA, P. W., FRENCH, C. J., and GUPPY, M. 1978. When and how the alpha-glycerophosphate cycle works. In Landry, F., and Orban, W. A. R., eds., *Third International Symposium on Biochemistry of Exercise.* Miami: Symposia Specialists. Pp. 29–42.

HOCHACHKA, P. W., and GUPPY, M. 1977. Variations on a theme by Embden, Meyerhof, and Parnas. In Jobsis, F. F., ed., *Oxygen and Physiological Function.* Dallas: Professional Information Library. Pp. 292–310.

HOCHACHKA, P. W., HULBERT, W. C., and GUPPY, M. 1978. The tuna power plant and furnace. In Sharp, G., and Dizon, A., eds., *The Physiological Ecology of Tunas.* New York: Academic Press. Pp. 153–174.

HULBERT, W. C., GUPPY, M., MURPHY, B., and HOCHACHKA, P. W. 1979. Metabolic sources of heat and power in tuna. I. Muscle fine structure. *J. Exp. Biol.*, 82: 289–301.

TAEGTMEYER, H. 1978. Metabolic responses to cardiac hypoxia: increased production of succinate by rabbit papillary muscles. *Circ. Res.* 45:808–815.

VAN DOP, C., HUTSON, S. M., and LARDY, H. A. 1977. Pyruvate metabolism in bovine epididymal spermatozoa. *J. Biol. Chem.* 252:1303–1308.

WILLIAMSON, J. R., SAFER, B., LaNOUE, K. F., SMITH, C. M., and WALAJTYS, E. 1973. Mitochondrial-cytosolic interactions in cardiac tissue: role of the malate-aspartate cycle in the removal of glycolytic NADH from the cytosol. *Soc. Exp. Biol. Symp.* 27:241–281.

Chapter Seven

Integrative Mechanisms in Hypoxia-Adapted Fish

In order to explore fundamental but adjustable components in anaerobic metabolism, our discussion to this point has intentionally emphasized closed-system mechanisms utilized in extending anoxia tolerance. Of course, the experimental isolation of organs and tissues leads to enormous oversimplifications of real situations in nature. In the real world, organs and tissues do not function in isolation neither in normoxia nor during extremes of hypoxia. On the contrary, they remain in communication with each other and all of them therefore function more or less as open systems, raising the possibility of crucial metabolic interactions between organs and tissues in order to extend the anoxia tolerance of the organism as a whole. Are such metabolic interactions directly demonstrable? If not, are they implied from indirect evidence? If or when they occur, are they of functional significance, in which case the interactions could be defined as cooperative? A good place to begin the exploration of these questions is with a group of organisms whose physiology and metabolism is well tuned to sustaining prolonged periods of reduced or fully depleted O_2 supplies. Hypoxia-adapted fish are one such groups of organisms.

Natural historians have found that many species of fish seem to possess impressive capacities for tolerating anoxia or hypoxia. Frequently, the capacity correlates with environmental O_2 availability. The advantages are self-evident: the greater the anaerobic potential, the greater the survival chances in environments poor in, or devoid of, oxygen. This was the explanation implied, for example, when Mather (1967) demonstrated that *Rasbora*, a cyprinid fish common to small ponds, pools, ditches, and streams of India, could survive in a sealed jar for more than 100 days! That too is the reason given to explain why goldfish and carp have developed pronounced anaerobic capacities. The European carp often survive in small ponds or lakes that freeze over in the winter and through the winter months become totally anoxic. It is often reported that the condition of these fish at spring thaw is as good as, or better than, it was at freeze-up.

In some cases, the above correlation is not evident. Instead, a highly developed anaerobic metabolism (coupled with a poorly developed aerobic one) seems to be one consequence of a low energy requirement or a low energy turnover. If food, for example, is limiting, it is of advantage to make a given caloric intake go as long as possible; under these conditions a low aerobic metabolism may be selected. At the same time, the organism's anaerobic potential may be maintained (or even improved) for emergency functions (for example, predator escape, prey capture, reproductive activity). Thus, organisms that tick over slowly may display some features in common with those selected for survival in low O_2 conditions. Abyssal fish appear to fit this category.

Low Oxygen Uptake by Abyssal Fish

To many biological oceanographers, insight into the bioenergetics of abyssal organisms has been a major, almost utopian goal. Recently, important advances have been made on two fronts: direct measurements of O_2 uptake in situ, and indirect estimates of anaerobic/aerobic potentials from enzyme profiles. The enzyme studies show a sharp rise in the ratio of anaerobic/aerobic metabolism, while the direct measurements support these findings and indicate that the O_2 uptake of abyssal fish may be as low as $\frac{1}{20}$ to $\frac{1}{10}$ that of "standard" eurythermic fish. One way of explaining this observation is to assume that the low temperatures and high pressures of the abyss depress metabolism. To some extent, this must occur, but at best this is only a partial explanation. If we assume a Q_{10} of 2 for O_2 uptake, the respiration rates of abyssal fish would only be about $\frac{1}{2}$ that of trout in summer in Montana mountain streams; that is, there still is an order-of-magnitude difference between the two kinds of organisms. Besides, Antarctic and Arctic fishes living at even lower temperatures than abyssal fish (at $-1.9°$ versus $+2.3°C$), are nicely cold-adapted and respire at substantially higher rates than would be expected from Arrhenius considerations. Clearly then, low temperature is not an acceptable explanation for the low respiration rates of abyssal fish. But what about high pressure?

Previously, pressure has been assumed to be an important environmental factor only for those organisms containing an air chamber. More recent studies, however, have indicated that hydrostatic pressure has important biological effects. This is a very complex area and it may be that the disrupting effects of pressure even today are generally underestimated. Many chemical/biochemical processes in theory could be seriously disrupted by pressures in the 100 to 500 atm range. All it would take is one such supersensitive function or structure to profoundly influence abyssal organisms. Janaasch and his coworkers (1973, 1977) in fact have observed that pressures at a depth of one mile are almost fully inhibitory to bacterial development and growth. Presumably, it is some particularly pressure-sensitive process or processes such as ribosomal function that accounts for this growth inhibition of bacteria in the mud and water of the abyss. Be that as it may, teleost fish have been filmed and captured at depths well in excess of those found to be inhibitory to bacterial growth and metabolism in the abyssal Atlantic. From the best available estimates of pressure effects on enzymes and metabolism of abyssal fish, pressures in the biological range (100 to 500 atm) would influence metabolism by less than $\frac{1}{10}$ as much as would the temperature differential (20°C) between surface and abyssal waters. Thus at least tentatively it seems safe to assume that the dramatic effects of pressure on microbes does not apply to vertebrates and that high pressure (like low temperature) cannot adequately account for the low respiratory rates of abyssal fish.

The Advantage of "Ticking Over" Slowly

We would be less hasty in rejecting temperature/pressure contributions to the low respiratory rate of abyssal fish if another fairly obvious explanation were not available. Oceanographers and marine ecologists have emphasized a scarcity of food in much of the midwater, bathypelagic, and abyssal waters of the open oceans. These areas are in effect liquid Sarahas. Abyssal organisms rely on food supplies literally raining down upon them. Like conventional rainfalls, these events are random and unpredictable. It is fair therefore to assume that food is at a premium. Under these conditions the advantages of a high anaerobic/aerobic metabolic potential are obvious and are similar to those of estivation and hibernation. By minimizing their energy requirements, abyssal organisms might maximize the length of time they could get along on a given bolus of food falling down from the waters above. Maintaining a high glycolytic potential also is advantageous because it allows short bursts of energy perhaps needed in emergency situations.

Environments similar to the oceanic abyss are not found in most freshwater lakes, rivers, and streams, but analogous situations are sometimes found in large, deep lakes. In the deepest, Lake Baikal, an extensive benthic fauna is known, although the metabolic organization of these organisms has not been described. The benthic fish of Lake Tanganika also present an intriguing situation. Lake Tanganika is permanently stratified and therefore contains no O_2 below the thermocline. Yet Coulter (1967) has described at least ten species of benthic fish. These presumably have a highly developed anaerobic metabolism, or have evolved vertically migrating behaviors that are yet to be discovered.

There are of course many other hypoxia-adapted fish, and there may be other selective forces contributing to their anaerobic abilities. These few have been singled out to expose new and potentially valuable experimental organisms to the classical biochemist or physiologist interested in mechanisms of hypoxia tolerance. It is probable (although not yet established) that many if not all of the adaptive mechanisms utilized in fact may be standard vertebrate responses, but exaggerated so as to illuminate underlying operating principles. (This may be particularly important in unraveling auxiliary anaerobic mechanisms that may be poorly developed in "normal" vertebrates.) One such principle is that biochemical adjustments are last resort strategies, and the first lines of defense occur at higher levels of organization. Although these levels are not our primary concern, they will be briefly described.

Physiological Responses to Hypoxia

If a fish has a choice, its first response to anoxic waters (predictably enough) is to leave! But, if such conditions are inevitable and imposed (by nature or by an experimenter), the behavioral strategy of anoxic fish appears to be one of wait-and-see. The common goldfish is a typical example, and remains almost totally quiescent when in anoxic water, ventilating at a low rate, responding only sluggishly to external disturbance. This kind of behavioral response is coupled to standard physiological reflexes. One of the first physiological responses, as the oxygen tension of the water decreases, is to increase ventilation rate and/or ventilation volume. A net increase results in the volume of available O_2 at the respiratory surface (thereby enhancing O_2 diffusion). This adjustment is effective only until the increased energy costs of ventilating the gills outweigh benefits derived from increased energy production. At this point ventilation rate drops to low values in complete anoxia.

Coupled with the ventilatory response to hypoxia and extending into anoxia is a marked bradycardia. Cardiac output remains relatively stable, but blood pressure simultaneously increases, presumably because of an increase in peripheral resistance. These overall "reflexes" are thought to extend the transit time of the blood through the gills, thereby allowing greater O_2 transfer.

In summary, hypoxic fish activate three reflexes: hyperventilation, bradycardia, and redistribution of cardiac output. At least the last two of these probably extend into anoxia. This implies that at least initially some organs are favored and remain relatively well perfused, while peripheral organs are hypoperfused. The peripheral organs obviously may need to rely most on anaerobic metabolism, but even the favored (and hence perfused) organs in complete anoxia must be sustained by anaerobic mechanisms. Unfortunately these are not well understood. The best information available seems to be for goldfish and carp (two quite closely related species), and our discussion will center on them.

Anaerobic End Products in Goldfish and Carp

Although other sources may be utilized to some extent, current evidence clearly identifies glycogen as the primary storage form of carbon and energy in the anoxic goldfish. It is therefore expected and found that during anoxia at 4°C, liver and muscle glycogen levels are reduced (by $\frac{1}{2}$ and $\frac{2}{3}$, respectively, after five days), while blood glucose levels rise from 2 mM to 14 mM at the expense of liver glycogen. The size of the glycogen depot determines how long anoxia can be sustained. Winter fish have larger livers, store proportionately more glycogen, and thus can survive anoxia longer than can summer fish.

If anaerobic glycolysis were the sole metabolic fate of storage glucose and glycogen in the anoxic goldfish, then it is clear that sustained anoxia should be associated with a large lactate accumulation. But it is, in contrast, the relative dearth of lactate following anoxia, noted by Prosser over twenty years ago, that was initially one of the most perplexing aspects of this problem in goldfish, and that has served as a point of departure for most subsequent work.

Lactate as an End Product

In a recent study, five days of anoxia led to a lactate buildup in goldfish blood of only 20 mM. If we make the limit assumption that all tissues are approximately in equilibrium with the plasma pool of lactate, it implies a metabolic rate during anoxia of about 0.003 μmol ATP/gm/min. That would be only $\frac{1}{300}$ the rate of ATP turnover in mammalian brain and about $\frac{1}{15}$ the rate of ATP turnover in goldfish during aerobic resting metabolism. Therefore either something else is occurring during anoxia (that is, auxiliary anaerobic mechanisms which can be assessed by novel anaerobic end products) or anoxia leads to a metabolic depression that is most unusual among the lower vertebrates.

Although both possibilities may be involved, the latter as the only explanation seems rather unlikely. In one of the most extreme cases of metabolic depression known in lower vertebrates, that of the estivating lungfish, metabolic rates are reduced only to about $\frac{1}{5}$ of normal "standard" rates. Our intuitive guess, therefore, is that a metabolic depression down to $\frac{1}{15}$ of normal is probably beyond the capacity of the anoxic goldfish, and this seems supported by the few (calorimetric) measurements that are available. These show the metabolic rate of anoxic goldfish is depressed to about $\frac{1}{5}$ the basal rate. Clearly, glycolytic rates assumed from lactate accumulation cannot account for the sustained anoxic tolerance of goldfish, and therefore something else must be occurring. At least a part of that "something else" involves CO_2 and ammonia production, since both, unexpectedly, are known to be released during anoxia in fish.

CO_2 as an End Product

A number of observations provide quite convincing evidence for the anaerobic production of metabolic CO_2 in the goldfish and carp. First, goldfish are able to maintain an RQ value of about 2 for two weeks at 15% air saturation, the result of decreased O_2 consumption. It is unlikely that CO_2 production for such long time periods could arise from anything but organic precursors. Second, CO_2 production can be demonstrated in isolated carp gills and goldfish muscle using manometric methods to separate nonmetabolic CO_2 released from HCO_3^- and true metabolic CO_2 released from metabolite precursors. Third, CO_2 production by anoxic carp occurs at over 200 μmoles/100 g/hr, a result first obtained by Blazka (1958) which agrees well with more recent data obtained by Van den Thillart (1977) on whole goldfish.

Although the above data are highly provocative, the most convincing demonstration of anaerobic production of metabolic CO_2 comes from studies using ^{14}C-labeled substrates. Carbon dioxide production at 10°C in anoxic gold-fish using glucose-1-^{14}C was first monitored in a simple study in 1961 (Hochachka, 1961). This was assumed to be an experimental artifact, and therefore the problem lay dormant for fifteen years. Then Van den Thillart and Kesbeke found anaerobic CO_2 production from ^{14}C-U-glucose in goldfish at 20°C. Based on the relative decrease in glycogen stores and accumulation of lactate in parallel sets of anoxia experiments, they calculated that only about 50% of the glycogen mobilized appears as lactate. The rest of the glycogen mobilized or the rest of the lactate formed must have some other fate. Recent studies by Shoubridge confirm the anaerobic production of CO_2 from glucose and from lactate. More significantly, Shoubridge has established that an important fate of part of the lactate formed is oxidation in organs remote from sites of formation.

When goldfish are subjected to carbon monoxide poisoning and subsequently held under CO or N_2, they continue to oxidize ^{14}C-U-glucose slowly but, in contrast, $^{14}CO_2$ release from ^{14}C-U-lactate occurs at high rates. Since $^{14}CO_2$ production from ^{14}C-1-lactate occurs in anoxia at about 70% the normoxic oxidation rate while $^{14}CO_2$ is not released from ^{14}C-3-lactate, pyruvate dehydrogenase can be identified as the site of CO_2 release. In the pyruvate decarboxylation part of the reaction scheme it is the carboxyl (C-1) carbon that comes off as CO_2. The reaction sequence for partial lactate oxidation in anoxic goldfish therefore is lactate $\rightarrow$ pyruvate $\rightarrow$ acetyl CoA + CO_2.

From the work of Shoubridge it is evident that carbohydrate is indeed fermented to lactate in the goldfish during anoxia, but at least some of the lactate (up to about 70%) is oxidized through the pyruvate dehydrogenase reaction, a process that explains why anoxic goldfish and carp typically do not accumulate large amounts of lactate. Lactate formation and lactate oxidation probably occur in different organs. Whereas lactate formation may occur in many tissues of the goldfish body, tissue slice studies show that lactate oxidation is most active in red muscle.

Oxidation of lactate through the pyruvate dehydrogenase reaction leads to the formation of acetyl CoA, so the question of the subsequent metabolic fate of this compound is of pressing importance. According to current thoughts, the two most likely alternatives involve either deacylation to form acetate or further derivatives of acetate as anaerobic end products or chaing elongation with various lipids as anaerobic end products.

Acetate or Acetate Derivatives as End Products

Acetyl CoA conversion to acetate is catalyzed by acetate thiokinase. In mammalian tissues, acetate thiokinase requires AMP and pyrophosphate (PP$_i$) as cosubstrates but in some lower animals (insects) thiokinases are ADP-linked:

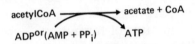

Hydrolysis of the thiolester bond proceeds with a large, negative free-energy drop, and it is tempting therefore to suggest that in anoxia the thiolester bond could be utilized for mitochondrial, substrate-level ATP synthesis. The use of acetate as an anaerobic end product could be viewed as beneficial on two other counts as well. First, because it is a weaker acid than is lactate (pK values of about 4.8 and 3.7, respectively), acetate as an anaerobic end product could constitute less of a buffering problem to the organism than would lactate. Secondly, acetate formation does not represent a metabolic cul de sac for acetate in theory could be reduced to ethanol:

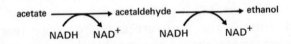

The enzymes catalyzing these reactions, acetaldehyde dehydrogenase and ethanol dehydrogenase, respectively, are best known in procaryotes, but are also present in mammalian liver. In the latter, they are usually thought to function in ethanol oxidation or in acetaldehyde conversion to acetate, the aldehyde being an extremely toxic metabolite. In anoxic goldfish, ethanol dehydrogenase occurs at ample activities in skeletal muscle, particularly in red fibers, and ethanol accumulates along with lactate and to similar levels. The above reactions thus appear to be utilized in the forward direction (acetate reduction to ethanol) in the O_2-deprived goldfish and the bulk of the ethanol formed is released to the outside water. However, despite the apparent advantages, in terms of energy yield and redox regulation, only preliminary data from the work of Shoubridge are available to suggest that these lactate-derived end products are formed (and excreted to the outside) during anoxic stress in the goldfish, and the quantitative contributions of these reaction schemes to energy metabolism in anoxic goldfish are not yet known.

Lipids as End Products

The idea of lipids as potential anaerobic end products is not really very new. Although not often measured in metabolic studies of fishes, it has been well established in other organisms. For example, the anaerobic incorporation of acetate and pyruvate into lipid in gastropods is comparable to aerobic incorporation. Moreover, ^{14}C from the end products of anaerobic glycolysis (alanine and lactate) also appears in fatty acids after anoxia.

Similar processes are well described in mammals as well. Whereat and coworkers (1967) demonstrated a system of fatty acid formation and chain elongation in rabbit heart mitochondria proceeding via a reversal of β-oxidation and apparently activated by high $NADH/NAD^+$ ratios and by succinate. In the isolated perfused rat heart, C^{14}-acetate incorporation into free fatty acids, triglyceride, and various phospholipids is greatly increased under anaerobic conditions. On average there is an eight-fold increase in labeling of neutral lipids and as much as a twenty-fold increase in labeling of phospholipids. In a nonrecirculating system, allowing the simultaneous measurement of lactate production and acetate incorporation, the anaerobic rate of acetate incorporation is about $\frac{1}{100}$ the rate of anerobic glycolysis. Therefore in the rat heart the process is probably irrelevant to overall anaerobic energy metabolism, even if it may play an important role in the maintenance of mitochondrial integrity in anoxia. In constrast, in hypoxic or anoxic fishes, distinct physiological advantages may accrue from accentuating this system, and its utilization has often been suggested in the literature. Definitive demonstrations of its contribution to anaerobic metabolism in fishes, however, are lacking. Nevertheless, the rat heart is not unique in showing enhanced fat deposition during anoxia. Of other tissues that show this capacity, the potential for anaerobic fatty acid formation is highest in liver, kidney cortex, and red muscle.

In mammals, controversy over the actual pathway of anaerobic fatty acid formation came to end in the last several years with the demonstration by several laboratories that the process basically involved chain elongation. Except for the final step, catalyzed by enoylCoA reductase, the pathway is essentially the reverse of β-oxidation. The free-energy change for the overall pathway (fig. 7.1) is about -10 kcal/mole, which makes the process readily feasible thermodynamically. The pathway in fact can be reconstructed with isolated enzymes, which supplies a final, convincing proof of its feasibility. In mammalian heart, it is a mitochondrial process and apparently the only means of fatty acid formation. In the liver, it also is localized to the mitochondria (although a similar system occurs in the microsomes), and in this tissue enoyl reductase utilizes NADPH in preference to NADH. Metabolites such as glutamate, which generate high NADPH/NADP$^+$ ratios, stimulate enoyl reductase in liver.

At least three functional advantages of forming lipids during hypoxia can be suggested. In the first place, the process regenerates NAD$^+$ within the mitochondria which can contribute to maintaining an oxidizing potential, particularly at the pyruvate dehydrogenase reaction. Another important function of the process is to conserve acetyl units in a storage form for reoxidation upon return to aerobic conditions. Lipids as anaerobic end products also are relatively innocuous; they do not lead to acid–base imbalances, they are light, and they are the most reduced of carbon substrate sources available to cell metabolism.

Although any or all of these may contribute to the se-

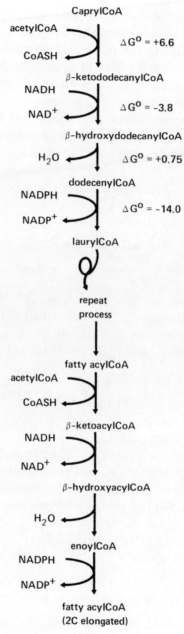

Figure 7.1 Mechanism of chain elongation during anoxia in mammalian heart. (After Seubert and Podak 1973.)

lective advantage of lipids as end products in hypoxia-adapted organisms, perhaps the biggest advantage stems from the possibility of coupling this process (particularly in the liver) with the catabolism of amino acids. The immediate source of excretory nitrogen in fishes is now generally believed to be glutamate, in a process catalyzed by glutamate dehydrogenase, which in the liver is $NADP^+$ dependent. Because of the large free-energy drop of enoyl reductase, it is thought the two enzymes form a redox couple:

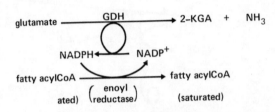

In mammals, this process (which in the isolated state is stimulated by glutamate) may feed into the urea cycle for the continued formation of urea, while in fish, it would explain the continued formation of ammonia during anoxia.

Ammonia as an End Product

Ammonia production in fish during anoxia in fact is relatively well established. In *Tilapia*, AQ ratios (volume NH_3 produced/volume of O_2 consumed) increase sharply when PO_2 falls below 2 ppm; the change in AQ results from decreasing O_2 consumption, but unchanging rates of NH_3 formation. Similarly, as observed by various workers, the rate of ammonia production in goldfish is unchanged between aerobic and anaerobic rates. From the ratio of NH_3/CO_2 production Van den Thillart (1977) calculated that 25% of anaerobically catabolized substrate in goldfish could be amino acid.

Role of Succinate and Alanine

From the above data it is tempting to view metabolism in the goldfish as a two-part system. In one part, represented by white muscle and other hypoperfused tissues, the main metabolic process occurring involves lactate formation. The second part of this system utilizes this lactate to form

CO_2 and other end products. In this view, glucose metabolism, particularly in red muscle, is supplanted by further lactate metabolism. Since pyruvate is an intermediate in the metabolism of both lactate and glucose, both processes require a source of NAD^+ to remain in redox balance. One possible way of balancing redox in the cytosol involves cytoplasmic MDH, with aspartate being the most likely source of OXA for the NADH-dependent formation of malate. Aspartate aminotransferase under such conditions in other organisms is usually linked to alanine aminotransferase (with glutamate and 2-ketoglutarate tumbling between them), and this too appears to occur in fish. Alanine in fish, as in mammals, is an important carrier of nitrogen to the liver, where in fish it is released as NH_3. Malate, on the other hand, does not appear to accumulate during anaerobiosis in fish, but it can be readily converted to succinate, and succinate is a well-established carbon sink in anoxic fish, as it is elsewhere. What is more, it has been found that succinate accumulates most in relatively aerobic tissues (such as the heart and red muscle), which would fit the above model. It is also of interest that succinate potently activates chain elongation in isolated systems.

It should be emphasized, however, that the quantitative contributions to anaerobic metabolism of these various processes cannot be currently specified; that they do play important roles is strongly indicated by the internal consistency of the available data. However, only future research can settle precise stoichiometric relationships, and when this is done, additional attention must be given to the potential role of amino acids.

Functional Advantages of the Goldfish Metabolic Organization

In summary, in addition to satisfying the usual requirements of anoxia metabolism (see chapter 1), the metabolic setup in the goldfish displays a few obvious advantages. Perhaps the most important of these is that it minimizes the buildup of lactate in the blood and tissues during anoxic excursions (fig. 7.2). This in turn prevents blood and tissue pH from falling to unacceptably low values or from falling at

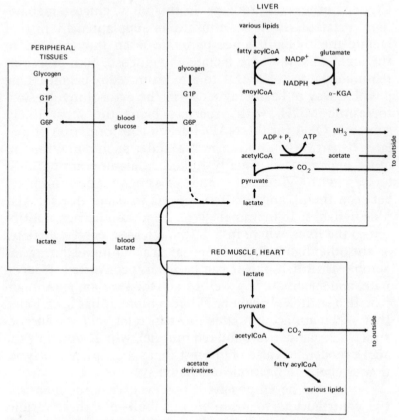

Figure 7.2 Summary metabolic model of hypoxia metabolism in the goldfish showing metabolic interactions between different tissues that minimize lactate accumulation. Carbon dioxide, ammonia, and ethanol (or some metabolic derivative of acetate) are released to the outside as waste anaerobic end products.

unacceptably high rates. Such a mechanism would be particularly advantageous in fish because they cannot maintain a large bicarbonate reserve in the blood; CO_2 equilibration across the gill is a very efficient process and the external medium serves as an excellent CO_2 sink.

Another important advantage of the goldfish metabolic organization is that a means is maintained for generating NH_3 during anoxia; without this process, amino acid fermentation would not be possible, nor would NH_3 be available for acid–base regulation. Moreover, at least three of the currently identified anaerobic end products (CO_2, ethanol, and ammonia) are all readily removed to the outside water and thus need not be accumulated internally to detrimentally high levels. Such obvious advantages, however, depend upon incorporating an element of cooperativity in metabolic interactions between different organs and tissues, which undoubtedly contributes to extending the anoxia tolerance of the organism as a whole.

Suggested Readings

Blazka, P. 1958. The anaerobic metabolism of fish. *Physiol. Zool.* 31:117–128.

Coulter, G. W. 1967. Low apparent oxygen requirements of deep water fishes in Lake Tanganyika. *Nature* 215:317–318.

Gloster, J., and Harris, P. 1972. Effect of anaerobiosis on the incorporation of ^{14}C-acetate into lipid in the perfused rat heart. *J. Mol. Cell. Cardiol.* 4:213–228.

Hinsch, W., Klages, C., and Seubert, W. 1976. On the mechanism of malonylCoA-independent fatty acid synthesis: different properties of the mitochondrial chain elongation and enoyl-CoA reductase in various tissues. *Eur. J. Biochem.* 64:45–55.

Hochachka, P. W. 1961. Glucose and acetate metabolism in fish. *Can. J. Biochem. Physiol.* 39:1937–1941.

Hochachka, P. W., ed. 1975. Biochemistry at depth. *Comp. Biochem. Physiol.* (B) 52:1–202.

Jannasch, H. W., and Wirsen, C. O. 1973. Deep sea microorganisms: *in situ* response to nutrient enrichment. *Science* 180:641–643.

———. 1977. Microbial life in the deep sea. *Sci. Am.* 236:45–52.

Kutty, M. N. 1968. Respiratory quotients in goldfish and rainbow trout. *J. Fish. Res. Board Canada* 25:1689–1728.

———. 1972. Respiratory quotient and ammonia excretion in *Tilapia mossambica. Marine Biology* 16:126–133.

Low, P. S., and Somero, G. N. 1976. Adaptation of muscle pyruvate kinases to environmental temperatures and pressures. *J. Exp. Zool.* 198:1–12.

Mathur, G. B. 1967. Anaerobic respiration in a cyprinoid fish *Rasbora daniconius* (Ham.). *Nature* (London) 214:318–319.

Prosser, C. L., Barr L. M., Pinc, R. A., and Lauer, C. Y. 1957. Acclimation of goldfish to low concentrations of oxygen. *Physiol. Zool.* 30:137–141.

Randall, D. J. 1970. Gas exchange in fish. In Hoar, W. S., and Randall, D. J., eds., *Fish Physiology*, vol. 4. New York: Academic Press. Pp. 253–292.

Seubert, W., and Podack, E. R. 1973. Mechanisms and physiological roles of fatty acid chain elongation in microsomes and mitochondria. *Mol. Cell. Biochem.* 1:29–40.

Shoubridge, E., and Hochachka, P. W. 1979. Lactate oxidation in the anoxic goldfish. *Intl. Congress. Biochem.* 13:1–R123.

Somero, G. N., Siebenaller, J. F., and Hochachka, P. W. 1979. Biochemical and physiological adaptations of deep-sea animals. In Rowe, G. T., ed., *The Sea: Deep-Sea Biology.* New York: Wiley.

Van den Thillart, G. E. E. J. M. 1977. In *Influence of oxygen availability on the energy metabolism of goldfish, Carassius auratus L.* Ph. D. thesis, State Univ. of Leiden, The Netherlands.

Whereat, A. F., Hull, F. E., and Orishimo, M. W. 1967. The role of succinate in the regulation of fatty acid synthesis by heart mitochondria. *J. Biol. Chem.* 242:4013–4022.

Chapter Eight
Air-Breathing Fish

In many tropical waters, annual flooding of the rain forest, high water temperatures, and high rates of O_2 utilization can lead to serious limitiation of molecular oxygen. Of several adaptive strategies displayed by fish in this region (skimming, simple increase in hypoxia tolerance, and so on), air breathing is one of the most interesting solutions to this critical problem of O_2 shortage in water. The metabolic advantages of breathing air, however, are procured at considerable cost. The most serious limitations imposed on aquatic organisms that turn to breathing air stem from being tied to an easily defined, two-dimensional space: the surface. Many ecological consequences follow, such as access to a smaller fraction of the water column, access to less food, possibly restricted breeding areas, fewer options for development and growth of juvenile stages, and, most importantly, greater exposure to predation particularly by aerial predators. For all or a combination of such reasons, it is probably of advantage to develop capacities for making a breath "go" as long as possible; as in other diving animals, such capacities may not be routinely utilized, but they would be of obvious advantage in emergency situations. Hence, air-breathing fish may be expected, and in fact are known, to display rather striking anaerobic capabilities.

In the previous chapter emphasis was placed on the idea of cooperative metabolic interactions between tissues varying in metabolic organization and capacities. According to such concepts, the whole organism in hypoxia may be divided into interacting functional units. With respect to glucose metabolism, some of these (including tissues such as skeletal white muscle) ferment glycogen (or glucose) and generate lactate; other units, which serve as metabolic "partners" of the first, including the liver, red muscle, and probably heart, utilize lactate in a metabolism that generates other anaerobic end products (CO_2, ethanol, and possibly lipids) and sets the stage for the continued production of NH_4^+ released during amino acid metabolism. The integrated process, which helps to explain the outstanding anoxia tolerance of species such as goldfish and carp, depends primarily upon the tissues of different functional units remaining in communication with each other. That is, each organ or tissue component of one functional unit remains an open system, in communication with other components. But there is no requirement for tight integration of these functional units in time. For at least one group of fish, the air breathers, such a requirement arises because any cooperative metabolic interactions between different organs must be timed with respect to apnea and breathing cycles. This is achieved through the refinement of physiological responses to breath-hold diving that can be loosely termed diving reflexes. Their nature and magnitude in air-breathing fish are perhaps best illustrated by the lungfishes.

Circulatory System of the Lungfish

The pattern of cardiovascular adjustments during breath-hold diving in the lungfishes is strongly determined by cardiovascular anatomy. The circulatory system of the African lungfish is shown schematically in figure 8.1. The branchial circulation differs in at least two important aspects from branchial circulation in other fish. First, the vascular exchange area represented by the primary and secondary gill filaments is much reduced; and second, the branchial circulation is separated into pulmonary versus systemic vascular beds. The branchial differentiation is coupled to a partial separation of blood flow through the heart. Thus, arches, I, II, and III receive oxygenated blood from the left side of the heart, delivering it to the dorsal

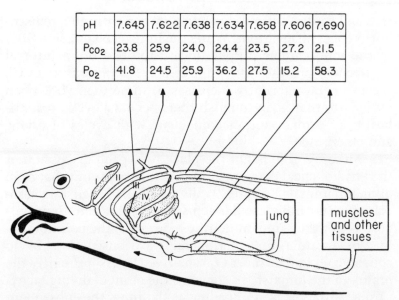

pH	7.645	7.622	7.638	7.634	7.658	7.606	7.690
P_{CO_2}	23.8	25.9	24.0	24.4	23.5	27.2	21.5
P_{O_2}	41.8	24.5	25.9	36.2	27.5	15.2	58.3

Figure 8.1 Schematic representation of the branchial circulation separated into pulmonary and systemic beds. The pH, PCO_2, and PO_2 values empirically demonstrate the partial separation of blood flow through the heart. (After Szidon et al. 1969.)

aorta and ultimately to the general systemic circulation. Arches IV, V, and VI receive deoxygenated blood from the right side of the heart and deliver it to the lungs. But because blood flow to the lung fluctuates through the breathing cycle, there is a need for an alternative pathway by which blood can return from the gills to the systemic circulation. This pathway is supplied by a ductus which is located at the origin of the pulmonary arteries and allows a shunt between arches IV-VI and the ascending aortic branches.

In *Lepidosiren* and its African relative *Protopterus*, the branchial vascular resistance is very low, undoubtedly because the large-bore, low-resistance vascular shunts through arches II and III have lost their gill filaments. This vascular bed as a result always gets a large fraction of the cardiac output. But although large, it is by no means all of the output. In *Protopterus*, blood flow increases four-fold in the pulmonary circuit in association with each air breath. The increase is in part due to a general elevation of heart rate and cardiac output, but more importantly to a redistribution of cardiac output. While the lung receives about 60% of the total cardiac output toward the end of an interval between air breaths (4 to 5 minutes in *Protopterus* at 25°C), the percentage abruptly increases to more than 90% upon air breath; that is, in lungfish there occurs a cyclic redistribution of cardiac output coincident with cycles of diving and emergence.

Although this is not the place to expand discussion of the physiological data, it is important to point out that as in diving mammals (chapter 9), diving in the lungfish is associated with cardiovascular adjustments, but here their purpose is not the same. In marine mammals, the point of diving bradycardia and peripheral vasoconstriction is to conserve O_2 for the most O_2-sensitive tissues (primarily the brain). In the lungfish, in contrast, the point of diving tachycardia is to increase O_2 loading at the lung; the subsequent drop in cardiac rate and output probably serve the same functions as in diving mammals: they extend the apneic period by limiting the rate of O_2 utilization by peripheral tissues.

In metabolic terms, an important implication of these cardiovascular structural constraints is that the heart is on a "most-favored" circuit, since it is the only major organ that is directly served by the pulmonary bed circuitry and which also forms a key part of that circulation. The brain in most vertebrate species also enjoys a rich perfusion even under hypoxic conditions, but this may not be the case in air-breathing fishes such as *Lepidosiren*. Since the brain is supplied by an *efferent* branchial system arising in arch I that in anoxic waters can actually contain *less* O_2 than arrives at the gill (because of loss to the water during lamellar perfusion), this organ may be uniquely limited in O_2 supplies, particularly in extreme situations (for example, in fully anoxic waters or while burrowing in mud).

In complete agreement with these impressions are enzyme and ultrastructural studies showing the heart as an oxidative "hot spot," with far higher capacities for oxidative metabolism than those found in any other organ or tissue in the body. Moreover the enzymatic potentials of various organs examined in air-breathing fish allow the construction of a model of close metabolic interactions between different organs during diving and recovery cycles. With respect to glucose metabolism during diving, the model emphasizes that some organs are predominantly glycolytic and generate lactate, while others utilize it. Because of its relative size, constituting about $\frac{3}{4}$ of the animal, the myotomal muscle must be a major site of anaerobic formation of lactate, yet many other organs (including the gills, the liver, the gastrointestinal tract, some kidney tissues, and even the brain also display high glycolytic, but low oxidative, capacities. On the other hand, the heart and distal kidney tubules appear fully capable of utilizing the lactate formed, a process that probably occurs mainly during the breath-hold period but also during recovery from any unusually large lactate loads. In the heart, the main fate of lactate appears to be complete oxidation (if O_2 supplies are adequate), while in the kidney, the predominant metabolic fate of lactate appears to be conversion to glucose. Such cooperative metabolic interactions are advantageous for they diminish the accumulation of lactate during the breath-hold period and may speed up the clearance during recovery of any accumulated lactate. As in hypoxia-adapted goldfish, the centerpiece of this model of metabolism during diving and recovery cycles in air-breathing fishes is the integration of activities in different organs and tissues of the body. These tissue-specific capacities, however, are not easily identifiable because of the complexities of organ metabolism and hence must be discussed in some detail. However, the arising model of metabolism in these fascinating animals is heuristic, predictive, and eminently testable, and these features should justify the burden of piecing together the puzzle. Although metabolite data in air-breathing fishes are sparse, some ultrastructural and enzyme studies of various organs are available for at least three good air breathers— *Arapaima, Lepidosiren,* and *Synbranchus*—and these will serve as the bases for discussion.

Brain Enzyme Adjustments

Of all organs in the vertebrate body, the brain is usually considered to be the most O_2-dependent. This view has been held for two reasons. First, early data showed that hypoxia or ischemia led to irreversible damage to "vulnerable areas" of the CNS. Whereas these data are not in dispute, more recent studies indicate that such partial stress may be more damaging than is total anoxia; thus this reason for expecting a highly O_2-dependent brain metabolism in all species is no longer as compelling.

The second reason is more complex. The mammalian brain normally displays a remarkable constant and high rate of ATP turnover. Although the brain may be only 2% of the body weight, it can account for 20% of basal metabolism (see chapter 9). Obviously, only short interruptions of O_2 or substrate delivery would therefore lead to potentially catastrophic curtailment of CNS function. In this view, an important reason for the extreme sensitivity of the brain to a lack of oxygen arises from the inability of glycolysis to generate ATP at a high enough rate. Furthermore, even though species with a high glycolytic potential also show a greater brain anoxia tolerance, in none thus far studied with the possible exception of the diving turtle *Pseudemys*, is glycolysis active enough to completely sustain brain function. However, this capacity may be present in *Arapaima*, a conclusion deriving from studies of brain enzyme levels (table 8.1).

In *Arapaima* brain, the levels of glycolytic enzymes are relatively similar to those found in other vertebrates. Phosphofructokinase (PFK) occurs at somewhat higher levels than in the ox brain, for example, while pyruvate kinase (PK) occurs at lower levels (probably because fish PKs are usually regulatory enzymes). Lactate dehydrogenase (LDH) activities per gm wet weight of tissue are almost identical when corrected for differing temperatures. However, unlike mammalian brain LDH, which is composed predominantly of H-type subunits and displays distinct bifunctional LDH activity, *Arapaima* LDH does not appear to be sensitive to pyruvate inhibition. In this kinetic characteristic it resembles the M-type LDH in mammals which is specialized for pyruvate reductase function.

These data of themselves are not surprising. What is surprising is the low absolute level of oxidative enzyme activities in *Arapaima* brain. Citrate synthase, a key control

Table 8.1 The activities of selected enzymes in brain energy metabolism of *Arapaima*. All activities are in μmol/min/gm at pH 7.0, 25°C.

Enzyme	Activity
Anaerobic metabolism	
Pyruvate kinase	51
Lactate dehydrogenase (low pyruvate)	59
Lactate dehydrogenase (high pyruvate)	56
Aerobic metabolism	
Citrate synthase	0.4
Glutamate dehydrogenase	1.5
Aspartate aminotransferase	20.9
Mixed functions	
Malate dehydrogenase	76
Glucose-6-P dehydrogenase	1.6
Phosphofructokinase	6
Fructose biphosphatase	0.3
Aldolase	7.0
α-glycerophosphate dehydrogenase	0.4
AMP deaminase	0.03

Source: Hochachka (1979).

site in the entry of acetyl CoA into the Krebs cycle, occurs at about $\frac{1}{20}$ to $\frac{1}{40}$ the levels found in the brain of a variety of other vertebrates including representative fish, reptiles, birds, and mammals. The only species in which similarly low levels of brain citrate synthase have been recorded is in an extremely hypoxia-adapted osteoglossid (aruana).

Thus we are left with the impression either that brain metabolic requirements in *Arapaima* are lower than in other vertebrates or that they are simply met by anaerobic metabolism. As mentioned above, because of anatomical constraints, this situation might be anticipated in the lungfishes as well. In both groups, the observed enzymatic potentials are presumed to represent part of the overall mechanism of O_2 conservation for more O_2-needy tissues, a metabolic strategy even more clearly expressed in *Arapaima* in the metabolic organization of skeletal muscles.

Arapaima White and Red Muscle

As in all fish, the osteoglossid myotome singly is by far the largest organ in the animal's body. In *Arapaima* essentially all of the myotome consists of large-diameter white fibers, which underlie small superficial lateral and small superficial dorsal bands of red fibers. As may be expected, *Arapaima* utilizes burst-type swimming.

From ultrastructural studies, the striking feature of *Arapaima* white muscle is its evident lack of oxidative capacity. Capillarity is extremely low, and lipid is absent within and between the fibers. Mitochondrial content is so low that scanning of many electron micrographs is necessary for observing them. The mitochondria are small in size and all peripheral in location.

The myotome of *Arapaima* contains only small amounts of red muscle, and its ultrastructure also implies a high glycolytic capacity. No intracellular or extracellular lipid is discernible, but numerous glycogen granules are present. Mitochondria, even if far more abundant than in white muscle, occupy only 3 to 5% of the area of electron micrographs; thus they are by no means as numerous as in more active species such as the tuna, where they occupy 35 to 50% of the area of similar micrographs. Red muscle in other teleosts appears intermediate to *Arapaima* and tuna.

The impressions gained from electron microscopy are that *Arapaima* white muscle is extremely anaerobic compared to red, and *Arapaima* red muscle may be relatively more glycolytic than most teleost red muscles. These impressions are further clarified when enzyme profiles of the two muscle types are compared (table 8.2). Such profiles of *Arapaima* white and red muscle are surprising in that catalytic activities of several enzymes in anaerobic glycolysis (including PFK, PK, LDH, and aldolase) occur at essentially the same levels in both muscle types. This is unusual when compared to other teleosts, and implies a high ratio of glycolytic-oxidative capacities in both tissues.

Malate dehydrogenase (MDH) activity is particularly

Table 8.2 The activities of selected enzymes in energy metabolism of *Arapaima* myotomal muscles at 25°C, pH 7.0. Average activities expressed as μmoles substrate converted to product/min/gm wet weight.

Enzyme	Activity	
	White muscle	Red muscle
Anaerobic metabolism		
Pyruvate kinase	103	134
Lactate dehydrogenase (low pyruvate)	260	263
Lactate dehydrogenase (high pyruvate)	182	181
Aerobic metabolism		
Citrate synthase	1.7	3.3
Glutamate dehydrogenase	1.3	3.0
Aspartate aminotransferase	11.2	54.4
Mixed functions		
Malate dehydrogenase	140	221
Glucose 6-P dehydrogenase	0.4	1.7
Phosphofructokinase	12.3	9.2
Fructose biphosphatase	0.7	1.0
Aldolase	19.8	21.4
α-glycerophosphate dehydrogenase	7.9	10.2
AMP deaminase	0.57	0.7

Source: Hochachka (1979).

interesting since this enzyme may play roles in both anaerobic and oxidative metabolism. In *Arapaima* MDH activity is somewhat higher in red muscle than in white. Since mitochondria are not very abundant in either muscle type, the bulk of the MDH in both must represent activity of the cytoplasmic isozyme. As in other systems, cytoplasmic MDH presumably takes on an auxiliary anaerobic function in the balancing of redox during specific stages of glycolytic activation. Glutamate-oxaloacetate transaminase (GOT) also occurs in both cytoplasmic and isozymic forms and its relatively high activity in both red and white muscle may in part correlate with MDH activities. In red muscle, however, the increased occurrence of mitochondria implies that both MDH and GOT also play roles in hydrogen shuttling during aerobic glycolysis.

In contrast to the above enzymes, most of which contribute in some way to anaerobic metabolism, the activities of enzymes functioning more restrictively in oxidative metabolism are higher in *Arapaima* red muscle than in white muscle. This would be predicted from the greater abundance of mitochondria in red muscle. These values can be viewed in better perspective, however, if compared to those of more active species; citrate synthase in tuna white muscle occurs at three- to four-fold higher levels than in *Arapaima*, while tuna red muscle citrate synthase is nearly ten-fold higher than in *Arapaima* red muscle. At the same time, the absolute mass of red muscle in tuna compared to the total locomotory muscle mass is of course much higher than in *Arapaima*. Similarly, glutamate dehydrogenase (GDH) activities in tuna red and white muscle are about two to three times higher than in *Arapaima*.

It is therefore difficult to avoid the impression that even if *Arapaima* breathes air, it nevertheless sustains an overall dampening of oxidative potential in both white and red muscles. Since the myotome is such a large proportion of the overall weight of these organisms, the enzyme measurements imply that the overall muscle contribution to O_2 consumption in *Arapaima* may be lower than in more active species. The reason for this outcome is not difficult to understand for, as in many diving vertebrates, in *Arapaima* it probably is useful to be able to extend the breath-hold period. By maintaining a relatively low oxidative capacity in the largest organ in the body (the myotome), O_2 taken in at the lung is automatically conserved for a longer time or for other more O_2-dependent tissues and organs. In consequence, lactate must be accumulated in muscle, and the question of its ultimate metabolic fate arises. As I shall discuss below, this lactate is probably utilized by other tissues either during the breath-hold period or during (O_2-rich) recovery periods, a process that would be advantageous since it would diminish the lactate load the organism must carry. Indeed, the ultrastructural and enzyme studies indicate that

lactate formed in muscle would have to be further metabolized elsewhere since in situ oxidative capacities are much
reduced. This metabolic strategy is even more pronounced
in *Lepidosiren*, the South American lungfish.

Lungfish White and Red Muscle

Again, the ultrastructure of lungfish white muscle is
similar to that in other fishes. Lipid is absent between and
within sarcomeres and mitochondria are extremely rare;
when present, they are almost always peripheral in location.
Capillary density is low, one capillary serving about fifteen
fibers. An extreme contrast to this is tuna white muscle,
with one capillary per fiber. These indications of a white
muscle largely sustained by anaerobic metabolism are also
implied by unusual glycogen storage mechanisms.

To fully appreciate this observation it is important to
recall that although fish white muscle is generally considered to be strongly dependent upon anaerobic glycolysis, it
paradoxically stores relatively small amounts of glycogen,
typically less than does red muscle. In contrast to this picture, lungfish white muscle stores an abundance of glycogen, located primarily between fibers, but also in myofibrillar regions such as near Z-lines or at myofibrillar branching
regions. Unlike the β-granules normally found in vertebrate
muscle, large diameter (750 to 900 Å α-particles or glycogen
rosettes are stored in lungfish white muscle (fig. 8.2). These
glycogen rosettes appear to represent a more efficient way of
packaging large quantities of glycogen for they are a relatively common storage form in the liver or in other tissues
such as the kidney; this is especially evident in certain
pathological conditions favouring extremely high glycogen
deposition. Apparently fused β-particles of glycogen (resem-
bling α-particles) have been seen in vertebrate muscle previously, but only in exceptional circumstances when they
are organized into distinct glycogen bodies. Thus the occurrence of α-particles in lungfish white muscle as a major
form of glycogen deposition represents a completley unique
situation that is probably of advantage to the organism in
hypoxic episodes.

Fish red muscle usually displays higher quantities of

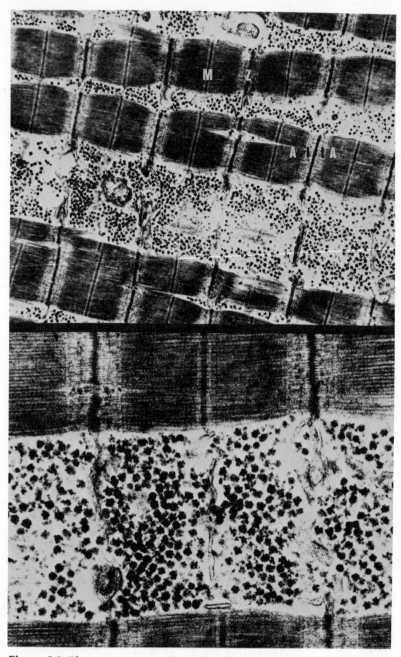

Figure 8.2 Electron micrograph of lungfish white muscle (×12,830 in upper panel and ×33,320 in lower panel), showing the M, Z, A, and I bands and the abundant α-glycogen particles (or rosettes) which range between 750 and 900 A in diameter. (From Hochachka and Hulbert 1978.)

glycogen than does white muscle; but in lungfish red muscle, what is again strikingly different is the way glycogen is stored, in this case either as glycogen bodies or as vast groupings of glycogen bodies and granules forming glycogen seas.

Glycogen bodies are usually defined as systematically arrayed glycogen-membrane complexes. They have been observed only periodically in a variety of vertebrate tissues, but they are very common in lungfish red muscle (fig. 8.3). Glycogen bodies in lungfish red muscle are both myofibrillar and peripheral in location. In either position in the muscle cell, they usually occur in close association with mitochondria, which are more abundant than in white muscle, as may be expected for a better perfused muscle (in light microscopy scans, on average one capillary per fiber is observed).

Lungfish red muscle also shows peripheral accumulations of glycogen granules which are so numerous and so uniformly dispersed that they are termed glycogen seas. They contain randomly dispersed mitochondria, usually small and round in shape, and at least in part are composed of aggregations of glycogen bodies. Such seas of glycogen are not randomly dispersed, but rather appear to occur as periodic pockets, usually sequestered near to the large nuclei that characterize this tissue. In other species, they also occur in subsarcolemmal positions. In view of their absence around the contractile elements of lungfish red muscle, it is probably that glycogen seas represent a specialized emergency reservoir that would not be utilized in normal activity patterns. That perhaps is why the glycogen seas show little evidence of depletion under conditions leading to extensive depletion of normally distributed glycogen.

Large glycogen reservoirs would be most useful to a relatively anaerobic muscle; they would seem to be less important in typical red muscle, which is considered to be a relatively aerobic, O_2-dependent tissue. Interestingly, lungfish red muscle does not display features necessary for a highly oxidative tissue. Capillarity is low by teleost standards for red muscle, about one capillary per fiber, which is more like tuna white muscle. In tuna red muscle, by con-

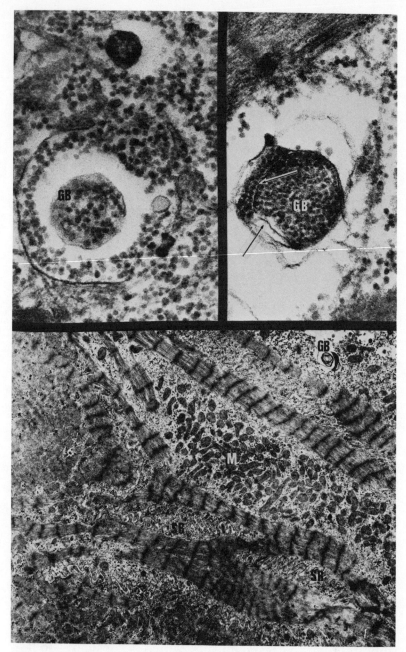

Figure 8.3 (*Upper panel, left*) Electron micrograph of lungfish red muscle (×62,240) showing a partially depleted glycogen body. (*Upper panel, right*) A similar electron micrograph (×70,770) with enough of the internal glycogen granules present to show their highly ordered nature. (*Lower panel*) Electron micrograph of lungfish myocardium (×6,000), showing extensive development of interdigitating sarcoplasmic reticulum (SR). These form a series of interfibrillar cisternae of varied size or a series of canals of varied shape and length. Mitochondria are numerous and mainly myofibrillar in position. (From Hochachka and Hulbert 1978.)

trast, each fiber is surrounded by 4 to 12 capillaries. If the tuna red muscle represents an extreme case of high capillarity, lungfish red muscle approaches the opposite extreme. Associated with the low capillarity are low mitochondrial numbers compared to other teleosts. And finally, lungfish red muscle displays little or no intracellular fat. Fat is usually considered the best fuel available for oxidative metabolism in teleost red muscles; its absence in lungfish red muscle thus is consistent with an overall view of a relatively anaerobic metabolism.

The impression gained from the above is that the lungfish myotomal muscles (both red and white) have unexpected anaerobic capacities—unexpected because *Lepidosiren* is a capable air breather. One possibility is that these muscles have become adjusted to sustain a diving habit. Whereas this metabolic organization obviously can serve in such a role, the result compared to *Arapaima* seems in excess of the need. Something is therefore missing in this explanation and the selective advantage of this arrangement must relate to other factors. The two most likely factors are burrowing and estivation. Unlike *Arapaima*, which courses freely in the water column, the South American lungfish, *Lepidosiren*, digs and inhabits burrows in moist mud and, if the conditions are suitable, can survive dry periods by estivation. The African lungfish *Protopterus* is even more famous for the same capacity. In view of this lifestyle, it is to be expected that these lungfishes should be profoundly tolerant of O_2 lack even if they are also most effective in breathing air. Particularly during burrowing in mud, under effectively anoxic conditions, the unexpected quantity of glycogen, and a high capacity for fermenting it, might well sustain metabolism and work for periods of time that would be impossible for less well adapted species. If this were indeed the case, one might expect that even the heart may show impressive anaerobic capacities. On the other hand, if this were not the case, then the lungfish heart may have relatively normal ratios of oxidative/fermentative metabolism. Indeed, since the heart and lung are on a direct circuit under all conditions, the ratio of aerobic/anaerobic metabolism might even be unexpectedly high. In this context, it is therefore instructive to examine the heart, which along with the brain in vertebrates is usually considered the most O_2-dependent organ in the body.

Lungfish and Synbranchus Hearts

Many of the features of the lungfish heart are similar to those described by others in different species, but there are some apparent specializations. The most striking of these in the lungfish heart is the extensive development of a modified sarcoplasmic reticulum (fig. 8.4). Its most characteristic shape and position is interdigitating between fibrils, forming a series of interfibrillar cisternae of varied size or a series of canals of varied length; these are in close association

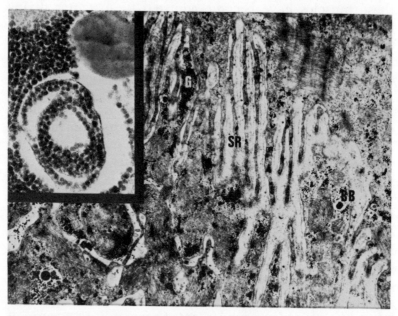

Figure 8.4 Lungfish myocardium ($\times 10,450$) showing a higher-magnification view of the extensive SR forming complex membrane-glycogen complexes. The empty canals are exactly reminiscent of similar membrane complexes in rabbit extraocular muscle when the glycogen granules are removed enzymatically. Some of the cisternae are glycogen filled (G). Periodic black bodies (BB) are evident; these are found in fully depleted (and collapsed) glycogen bodies, such as is shown in the inset ($\times 95,800$) in close association with a lipid droplet. (From Hochachka and Hulbert 1978.)

with glycogen and more rarely with mitochondria. In higher magnifications they can be observed to form parallel cisternal canals, often containing glycogen particles. This arrangement of the SR is highly reminiscent of evenly distended cisternal arrays in glycogen bodies of various mammalian tissues. In addition, packaged adjacent to this modified SR are the more usual type of glycogen bodies, already described in lungfish red muscle (fig. 8.4, inset). Glycogen granules vary in size in lungfish hearts (170 to 830 A diameter), the largest granules usually being associated with distended cisternal canals, while the smallest granules are those randomly dispersed.

Synbranchus, a South American eel-like fish, shares three key biological attributes with the lungfish: it is a capable air breather, it digs and inhabits burrows, and it is capable of estivation. In Synbranchus also heart ultrastructure is set off from that of other fishes by an extensive elaboration of the interfibrillar SR. As in Lepidosiren, glycogen granules and the interdigitating SR form an interfibrillar laminar structure in which empty and glycogen-filled cisternae of classical membrane-glycogen complexes are clearly evident.

All of these features of the synbranchid and lungfish hearts imply accentuated fermentative capacities. It is all the more interesting, therefore, that heart mitochondria are certainly abundant, adopting mainly oblong shapes, but also a variety of other forms. Under high magnification, the cristae are seen to be fairly thick and of medium electron opacity, while the matrix of the mitochondria is electron transparent.

The mitochondria are often associated with normal β-particles plus rounded glycogen bodies. However, as in "normal" vertebrate hearts, intracellular lipid is also available as a fuel for cardiac metabolism. Although present, lipid droplets do not appear as abundant as in other air breathers such as Arapaima. Nevertheless, the abundance of mitochondria plus intracellular lipid together imply a potentially important aerobic metabolism. In this connection, it should be emphasized that glycogen, in contrast to fat, of course can be mobilized either for oxidative or for fermentative metabolism; so from storage products alone it is not possible to decide which kind of metabolism is fueled by glycogen. However, an indication of why so much glycogen is stored and how it is used can be obtained from enzyme studies.

As expected from mitochondrial numbers, these hearts display high levels of enzymes in oxidative metabolism, such as citrate synthase, GDH, MDH, and aspartate amino-transferase (GOT). The occurrence of high activities of GOT and MDH but low α-glycerophosphate dehydrogenase (α-GPDH) activities, moreover, is consistent with oxygraph studies of cell-free heart preparations, showing that the malate–aspartate cycle is a more active hydrogen shuttling mechanism than is the α-glycerophosphate cycle. These results are clearly consistent with the electron microscopy in implying that when O_2 is available, the hearts of these air breathers are capable of a vigorous oxidative metabolism. The same is true in *Arapaima*, where heart oxidative metabolism can be fired by either fat or carbohydrate.

The observed levels of enzymes in oxidative metabolism are not far out of line with the levels of citrate synthase in the hearts of other fish. What is out of line with all other data on enzyme levels in fish hearts is the level of lactate dehydrogenase. Lactate dehydrogenase activity in the heart of both the lungfish and *Synbranchus* is in the order of 1000 μmoles substrate converted to product/min/gm wet weight of tissue (table 8.3). That is two to three times higher than found in the hearts of Amazon erythrinids and osteoglossids, and three-fold higher than found in the tuna heart. This LDH activity is three- to five-fold higher than is found in most mammalian hearts (with a temperature correction applied). The only other species showing comparably high heart LDH levels is the Antarctic Weddell seal, an outstanding pinniped diver, storing huge quantities of glycogen in the heart and capable of breath-hold diving for over an hour (see chapter 9).

Bifunctional Heart Lactate Dehydrogenases

An important functional characteristic of these potent LDHs in lungfish and *Synbranchus* hearts is high sensitiv-

Table 8.3 The activities of several selected enzymes in heart energy metabolism of three air breathing fishes at 25°C, pH 7.0. Activities are expressed as μmoles substrate converted to product/min/gm wet weight of tissue. Enzymes arranged according to known metabolic functions, though in some cases (for example, MDH) these may be overlapping.

	Activity		
Enzyme	*Lepidosiren*	*Synbranchus*	*Arapaima*
Redox regulation			
Lactate dehydrogenase (low pyruvate)	1180	870	370
Lactate dehydrogenase (high pyruvate)	465	385	245
Malate dehydrogenase	310	860	389
NADP⁺ isocitrate dehydrogenase	9	10	—
Hydrogen shuttling			
α-glycerophosphate dehydrogenase	0.2	0.4	0.5
Aspartate aminotransferase	40	77	49
Oxidative metabolism			
Citrate synthase	12	14	10
Glutamate dehydrogenase	3	9	4

Source: Hochachka and Hulbert (1978); Hochachka (1979).

ity to pyruvate inhibition. This identifies them as being composed predominantly of H-type subunits. H-type LDHs, as discussed in chapter 6, are characerically bifunctional; they are effective catalysts for either lactate production or lactate oxidation. But when are pyruvate reductase and lactate oxidase functions utilized? How is the bifunctional nature of these LDHs exploited?

To answer these questions, we must recall that the co-occurrence of extraordinary amounts of glycogen and of LDH in the hearts of the lungfish, *Synbranchus*, and the Weddell seal implies that in addition to a capable aerobic metabolism (comparable to that found in other vertebrates), these organs have developed powerful anaerobic glycolytic capacities. In the seal, such capacities clearly would serve as an emergency mechanism supporting heart work during prolonged diving. Similarly, one reason for the very high levels of heart LDH and heart glycogen in the lungfish and *Synbranchus* probably is to solve problems of periodic O_2 lack, in this case arising in them both during burrowing in the mud. Under these conditions, LDH would function as a pyruvate reductase and lactate would necessarily accumulate. However, in *Arapaima,* this situation probably would occur rarely, if at all, and LDH here would usually serve as a lactate oxidase.

As shown by Szidon and his colleagues (1969), air breathers share with the Weddell seal another common metabolic problem: that of handling a large "wash-out" of lactate from tissues such as muscle during recovery from anoxic or hypoxic episodes. The vertebrate heart is known to utilize lactate well under these circumstances, which is why, on kinetic criteria, heart LDHs in lungfish, *Synbranchus,* and *Arapaima* are composed predominantly of H-type subunits (tables 8.3 and 8.4). Although H-type LDHs are fully capable of bidirectional function, they are now widely believed to be kinetically better suited for function as lactate oxidases than is the muscle-type isozyme. Not surprisingly, O_2 uptake studies using cell free preparations from *Lepidosiren* and *Synbranchus* indicate that lactate, with NAD^+ present, is an excellent carbon and energy source for the heart. A qualitatively identical picture emerges from studies of cell-free preparations of *Arapaima* heart. So at least at the level of heart metabolism, these three air-breathing fish solve the problem of lactate loads in a manner rather similar to that observed in diving animals.

Table 8.4 Michaelis constants (mM) for heart and muscle lactate dehydrogenases in *Arapaima,* assayed at 30°C, pH 7.5.

Michaelis constant	Heart LDH	Muscle LDH
$K_{m(pyruvate)}$	0.15	1.0
$K_{m(lactate)}$	2.9	10.0

Source: Data from French and Hochachka (1978).

The Heart as an Oxidative Hot Spot

Although heart levels of oxidative enzymes in lungfish, *Synbranchus*, and *Arapaima* are similar, they are lower than in the hearts of more active species (tuna, dogfish, and trout) or in mammalian heart. Nevertheless, enzymes of aerobic metabolism in the hearts of air-breathing fish occur at higher activities than in any other organs sampled, including the brain. For example, in *Arapaima* heart, citrate synthase occurs at about 3, 5, 10, 10, 10, and 25 times higher levels than in red muscle, gill, white muscle, kidney, liver, and brain, respectively. Compared to other organs, the heart in these species is a veritable hot spot of aerobic metabolism, displaying enzyme levels that are about an order of magnitude higher than anywhere else in the body. Evidently, on a weight basis, the heart is capable of the highest sustained ATP turnover rates found in these species. By criteria of O_2 need, then, the heart must fit the category of a "favored" organ, for which O_2 supplies are conserved during diving. The only tissues that may fall into a similar category are the distal tubules of the kidney, which are loaded with mitochondria. In this case, the high O_2 dependence stems in part at least from an increased kidney role in ion and waste nitrogen metabolism. This can be illustrated in *Arapaima*, the only species in which kidney metabolic biochemistry has been studied.

Adjustments in Kidneys and Gills

In water-breathing fishes, the transport work load of the kidney is typically shared by the gills (which may in fact do the bulk of it), and although the nephron structure, with minor modifications, is similar to the mammalian homologue, the fish kidney does not make a large percentage contribution to ion regulation. In contrast, the transition to obligate air breathing in fishes is often associated with a decrease in the size and perfusion of the gills. For a given body size, the gill weight in *Arapaima* is about one half that in a related water breather (aruana), to which it may be conveniently compared.

In the absence of any other changes, this size difference would influence the exchange of materials across the gill,

but such differences between the two species are further accentuated by structural adjustments of the *Arapaima* gill,
allowing a circulatory shunt through the gill comparable to
that in the lungfish. Thus, perfusion of gill filaments can be
greatly reduced. Although the gills in *Arapaima* may still
play an important role in the release of CO_2, the reduction
in size and perfusion probably reduces their potential role in
ion regulation. In concert with this trend, kidney mass and
function are both expanded. Moreover, the enlarged kidney
is not divided into head and trunk regions; nephrons instead
are found throughout its length. This means that the increase in functional nephron mass is substantially greater
than implied by relative weight. Judging by the levels of
readily extractable enzymes involved in kidney function
(GDH being the most accurate index), the fractional contribution to NH_4^+ regulation by the kidney may be increased
by nearly an order of magnitude in *Arapaima*.

In terms of cell composition and ultrastructure, the
nephron in both species consists of three main segments.
The first segment is similar to that in other teleosts. Cells
of the second segment are more specialized, displaying a
highly infolded basilar membrane, lying in close laminar association with giant, elongate mitochondria, localized to
the basal $\frac{2}{3}$ of the cell. Cells of the third segment form the
longest homogeneous region of the *Arapaima* nephron. The
cells of the third segment are remarkably similar to teleost
chloride cells. Termed renal chloride cells, they are very
rich in mitochondria and are undoubtedly actively involved
in ion transport. It is the second and third segments that
probably set the minimal O_2 demands of the kidney. These
O_2 demands, of course, are also linked to the metabolic processes being performed. A better appreciation of these arises
from comparisons of the kinds and amounts of enzymes
present, that is, upon the metabolic potentials of the kidney
in *Arapaima* compared to its water-breathing relative,
aruana.

With respect to enzymes mobilizing substrate sources,
it is important to note that although the mammalian kidney
is able to utilize a number of substrate sources, lactate and
glutamine are of primary importance. A near absence of glu-

taminase in the osteoglossid kidney rules out a role for glutamine in NH_4^+ excretion and metabolism. But as in mammals, the enzymology of the *Arapaima* kidney is consistent with its utilization of lactate. Thus, the enzyme responsible for initiating lactate metabolism, lactate dehydrogenase, occurs in quite high activities in the *Arapaima* kidney (table 8.5), substantially higher in fact than in the liver, which is considered the main site of lactate utilization in mammals. In *Arapaima*, kidney, LDH is about twenty-fold higher than in the liver, while in aruana, it is about eight-fold higher. Moreover, kidney LDH in *Arapaima* occurs at nearly three-fold higher levels than in aruana. This point deserves some emphasis and further explanation, for LDH can also be utilized in anaerobic glycolysis in the formation of lactate, a role which may be much more important in aruana than in the *Arapaima*, not only in the kidney, but in all other organs compared as well. The best indication of the higher anaerobic glycolytic potential in aruana kidney is the six- to seven-fold higher levels of PK (table 8.5). PFK, another key regulatory enzyme in glycolysis, also occurs at three-fold higher levels in aruana than in *Arapaima* kidney. And finally, in this connection, higher amounts of glycogen are stored in second and third segments in aruana. Evidently, the overall glycolytic potential of aruana kidney is higher than in *Arapaima*, which is why we conclude that *Arapaima* kidney retains higher amounts of LDH not for sustaining higher glycolytic capacities but primarily for initiating lactate metabolism. Consistent with this interpretation is the occurrence in *Arapaima* kidney of a regulatory PK showing control characteristics similar to those in mammalian liver; these include cooperative substrate saturation kinetics, alanine plus ATP inhibition, and potent activation by fructose biphosphate. It is widely accepted that these control features work together to greatly dampen PK activity during periods of activated gluconeogenesis, thus allowing a net flow of lactate-derived pyruvate toward glucose.

In mammals, a significant fraction of the lactate taken up at the kidney may be reconverted to glucose, and several lines of evidence suggest a similar function for these organs

Table 8.5 The activities of selected enzymes in energy metabolism of kidney, liver, and gill in two osteoglossids, the obligate air breather aruana and the obligate water breather *Arapaima*. Average activities are expressed as μmoles substrate converted to product/min/gm wet weight at 25°C, pH 7.0, imidazole buffer. Enzymes are arranged according to known metabolic functions, although some of these (for example, LDH) may be overlapping.

Enzyme	Kidney		Liver		Gill	
	Aruana	*Arapaima*	Aruana	*Arapaima*	Aruana	*Arapaima*
Anaerobic metabolism						
Pyruvate kinase	205	31	12.7	9.5	50	30
Lactate dehydrogenase (low pyruvate)	141	332	20.9	17	69	52
Lactate dehydrogenase (high pyruvate)	102	313	24	13.1	50	36.5
Aerobic metabolism						
Malate dehydrogenase	461	173	88.5	281	53	69
Citrate synthase	1.3	0.9	0.6	1.2	1.4	1.95
Glutamate dehydrogenase	3.6	23.1	4.5	3.1	2.0	3.2
Aspartate aminotransferase	19.9	89.8	28.6	30.1	8.4	14.7
Other functions						
Hexokinase	1.8	1.7	Not assayed		0.5	0.5
Glucose 6-P dehydrogenase	5.3	4.3	6.8	4.8	3.6	6.4
Phosphofructokinase	2.0	0.68	1.71	0.45	1.5	0.57
Fructose biphosphatase	1.6	3.6	7.9	6.15	0.37	0.71
Aldolase	1.31	7.8	7.0	3.7	3.2	3.5
α-glycerophosphate dehydrogenase	1.7	2.7	1.2	7.8	0.56	0.71
AMP deaminase	0.25	0.75	0.11	0.056	1.0	0.48

Source: Hochachka (1979).

in *Arapaima*. The relative activities of fructose biphospha-tase (FBPase) and PFK in the osteoglossid kidneys supply an important insight into this problem. Fructose biphospha-tase, catalyzing the hydrolysis of fructose biphosphate (FBP), is considered obligatory to the *de novo* synthesis of glucose from lactate or from a variety of amino and keto acid precursors. Since FBPase is a soluble enzyme function-ing in the same cell compartment as PFK, the net direction of carbon flow through the F6P ↔ FBP interconversion depends upon the catalytic potential for the two reactions. In *Arapaima* kidney, the FBPase/PFK ratio is about 6, while in contrast, it is near unity in aruana kidney, because of re-duced levels of FBPase and increased levels of PFK (table 8.5). Since the liver is usually considered the chief glucon-eogenic organ, it is instructive to compare these ratios in the liver as well. In *Arapaima* liver, the FBPase/PFK activity ratio is about 12, compared to about 5 in aruana liver. This ratio can be used as an indication of the potential for carbon flow through this "futile" cycle toward glucose (or glyco-gen); it clearly suggests that the liver and kidney in *Ara-paima* and the liver in aruana have a high potential for this process. But both osteoglossids display low liver LDH activ-ities compared to the LDH activities in the kidney; hence, liver gluconeogenesis probably depends less upon lactate than upon other glucose precursors. In fact, the large kidney mass, the elevated kidney LDH but low hexokinase activi-ties, and the obviously copious perfusion in vivo all imply that the kidney in *Arapaima* has become a major organ for lactate uptake and gluconeogenesis, in these functions to some extent augmenting the role of the liver.

Glucose Metabolism during Diving and Recovery

It is tempting to summarize the above information in terms of a simple model, a *modus operandi* if you will, of metabolism through diving and recovery (apnea and breath-ing) cycles in these air-breathing fish. According to this model, during routine diving, glycogen and glucose are uti-lized by myotomal muscle, brain, and other tissues in a largely fermentative metabolism that releases lactate. In most cases, such as muscle and brain, this occurs automatically since the oxidative capacities are low while glycolytic capacities are high; however, it is possible that physiological mechanisms—vasoconstriction, for example—may also favor anaerobic metabolism in such tissues. As shown by Szidon and colleagues, lactate accumulated in these tissues during diving spills out into the blood and much of the lactate load therefore must be carried to other organs for further metabolism. The key organs involved appear to be the heart, kidneys, and liver. In the heart, conditions are suitable for lactate oxidation: LDH, showing a high affinity for lactate, occurs at high levels, thus allowing function in the thermodynamically "uphill" direction; mitochondria are abundant, oxidative enzyme activities are high, and O_2 supplies are not limiting. In the kidney and liver, conditions are suitable for gluconeogenesis and a large fraction of the lactate taken up probably is converted back to glucose. This organization appears metabolically sensible since it diminishes lactate accumulation during apnea (that is, during diving), regenerates glucose, and automatically conserves O_2 either for longer time periods or for more oxygen-needy tissues (figure 8.5). Or both. These metabolic interactions between different functional units in the organism would be necessarily abolished if O_2 were to become fully depleted. There is no evidence for this in *Arapaima*, but it probably occurs frequently in mud or burrows in the case of *Lepidosiren* or *Synbranchus*. At such time, even organs such as the heart must be sustained by fermentation and indeed conditions then are suitable for anaerobic glycogenolysis: heart glycogen stores are impressively large, glycolytic enzyme titres are high, and the catalytic potential for pyruvate reductase function of LDH is higher than anywhere else in the vertebrate world.

At the end of diving (in water or in mud), when blood O_2 stores are recharged, it is probable that much of the lactate accumulated in hypoperfused tissues such as muscle is washed out in a manner analogous to that seen in diving animals (chapter 9). Lactate oxidation in the heart and lactate-primed gluconeogenesis in the liver and kidney would be even further activated under these conditions. As it is advantageous to spend minimal time breathing, the main recovery period must occur in early stages of diving; at this

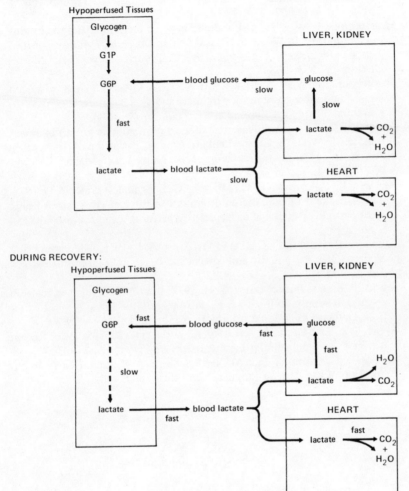

DURING DIVING:

DURING RECOVERY:

Figure 8.5 Schematic summary of probable metabolic events during diving and recovery cycles in the lungfish and *Arapaima.*

time lactate metabolism also must be most vigorous, for O_2 availability and lactate concentrations are at their highest.

Suggested Readings

FRENCH, C. J., and HOCHACHKA, P. W. 1978. Lactate dehydrogenase isozymes from heart and white muscle of water-breathing and air-breathing fish from the Amazon. *Can. J. Zool.* 56:769–773.

GUDERLEY, H., FIELDS, J. H. A., CARDENAS, J. M., and HOCHACHKA, P. W.

1978. Pyruvate kinase from the liver and kidney of *Arapaima gigas*. *Can. J. Zool.* 56:852–859.

HOCHACHKA, P. W. 1979. Cell metabolism, air breathing, and the origins of endothermy. In Wood, S. C., and Lenfant, C., eds., *Evolution of Respiratory Processes: A Comparative Approach*. New York: Marcel Dekker. Pp. 253–258.

HOCHACHKA, P. W., and HULBERT, W. C. 1978. Glycogen 'seas,' glycogen bodies, and glycogen granules in heart and skeletal muscle of two air-breathing, burrowing fishes. *Can. J. Zool.* 56:774–786.

JOHANSEN, K. 1979. Cardiovascular support of metabolic functiins in vertebrates. In Wood, S. C., and Lenfant, C., eds., *Evolution of Respiratory Processes: A Comparative Approach*. New York: Marcel Dekker. Pp. 107–192.

KRAMER, D. L., LINDSEY, C. C., MOODIE, G. E. E., and STEVENS, E. D. 1978. The fishes and the aquatic environment of the central Amazon basin, with particular reference to respiratory patterns. *Can. J. Zool.* 56:717–729.

LAURENT, P., DELANEY, R. G., and FISHMAN, A. P. 1978. The vasculature of the gills in the aquatic and aestivating lungfish (*Protopterus aethiopicus*). *J. Morphol.* 156:173–208.

SIESJO, B. K., and NORDSTROM, C. H. 1977. Brain metabolism in relation to oxygen supply. In Jobsis, F. F., ed., *Oxygen and Physiological Function*. Dallas: Professional Information Library. Pp. 459–479.

STOFF, J. S., EPSTEIN, F. H., NAIRNS, R., and RELMAN, A. S. 1976. Recent advances in renal tubular biochemistry. *Ann. Rev. Physiol.* 38:46–68.

SZIDON, J. P., LAHIRI, S. LEV, M., and FISHMAN, A. P. 1969. Heart and circulation of the African lungfish. *Circ. Res.* 25:23–38.

Chapter Nine

Diving Marine Mammals

The extension of anoxia tolerance through key metabolic interactions, integrated in time and between specific tissues, is nowhere better illustrated than in diving marine mammals. These processes must be so finely tuned because during prolonged diving the organism becomes a self-sustaining life-support system; isolated from its usual source of O_2, all necessary metabolic materials must be supplied from internal stores and thus must be "on board" when diving begins. Utilization of materials in short supply (O_2, blood glucose, and so on) must be regulated so that they may be conserved for the most needy organs or tissues. Metabolic end products must either be deposited or be more fully metabolized elsewhere. Satisfying such needs requires tissue-specific and readily regulated blends of aerobic and anaerobic metabolism. In long duration dives this is reflected by anaerobic metabolism sustaining work functions well before O_2 reserves of the blood are depleted. That is why during diving, even if blood O_2 supplies are still abundant, a large fraction (50% or more) of the blood glucose depleted is represented by a concomitant accumulation of lactate (table 9.1). Such a dependence upon anaerobic glycolysis during diving does not necessarily arise from a simple O_2 lack in the organism. Rather, it stems from the activation of an oxygen-conserving mechanism—the diving response—which leads to closely regulated partitioning of oxygenated blood between oxygen-needy and hypoxia-tolerant tissues.

145

Table 9.1 Relationship between the decrease in glucose concentration in systemic arterial blood and the rise in lactate concentration in the Weddell seal. In all cases, from half to all of the glucose utilized appears as blood lactate. As the seals were awake and varied in condition and degree of stress, no close correlation between diving time and glucose utilization is expected or evident.

Seal	Diving length (min)	Change in glucose concentration (μmol/ml)	Change in lactate concentration (μmol/ml)
1	5	−0.6	+0.5
2, Dive 1	10	−1.5	+1.3
Dive 2	22	−1.0	+1.7
3, Dive 1	20	−1.0	+1.1
Dive 2	25	−1.5	+1.5
4, Dive 1	26	−0.9	+0.8
Dive 2	46	−1.0	+2.5

Source: Hóchachka et al. (1977).

There are three components to the diving response: apnea, bradycardia, and peripheral vasoconstriction. Bradycardia leads to a drop in cardiac output, while peripheral vasoconstriction serves to maintain arterial blood pressure and redistribute blood flow preferentially to O_2-needy tissues. In order to be effective in conserving potentially limiting substrates (O_2, glucose), it is clear that most organs and tissues must be hypoperfused. Indeed, it was first anticipated that the function of the diving response was to conserve O_2 primarily for the heart, lung, and brain. The rest of the body was assumed to be largely hypoperfused. More recent studies, taking advantage of the outstanding diving abilities of seals, further clarify the situation. In terms of absolute perfusion rates, only the brain sustains a normal flow of blood and hence normal delivery of O_2 and carbon substrates; all other organs (with the exception of the adrenals, which remain quite well perfused) apparently experience a decrease in absolute flow rates, which may be consistent with the suggestion that the metabolic rate of the animal actually drops somewhat during diving. However, in terms of fractional cardiac output, four organs are identified as receiving an increased relative perfusion: the heart, lung, brain, and the adrenal glands. Tissues and organs such as skeletal muscle, liver, and kidneys sustain a sharp reduction both in absolute perfusion and in fractional cardiac output. The metabolic situation arising may be further complicated by partial compartmentation of the blood volume (fig. 9.1). In the Weddell seal, recent studies using labeled com-

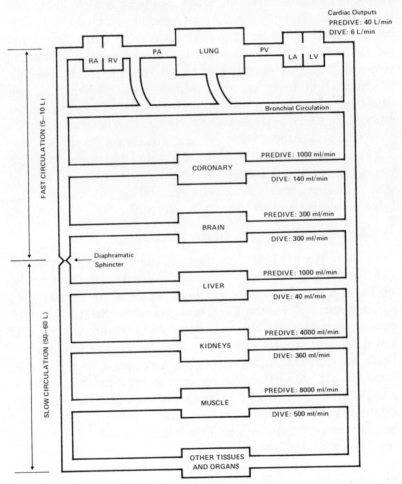

Figure 9.1 Schematic representation of circulatory adjustments during diving in the Weddell seal. Cardiac outputs and flow rates are rounded off and are representative of events at rest and during 8 to 12 minutes of diving. (From Murphy et al. 1980 and Zapol et al. 1979, with modifications.)

pounds estimate a "central" blood volume of about 8 liters that exchanges relatively slowly with the remaining 52 liter blood volume because of pooling in the venous system. For these reasons, a dichotomy (perfused versus hypoperfused tissues) is emphasized in the formal theory which in fact appears to extend into, and correlate with, enzymatic and metabolic adjustments.

The primary carbon and energy source utilized during diving is carbohydrate. Although glycogen may also be mobilized, the contribution of endogenous glycogen has not been quantified for any tissue nor any diving mammal thus far studied. However, at least for the Weddell seal during simulated diving, patterns of blood glucose levels through diving and recovery are now well defined. The unexpected finding is that during diving in the seal a part of the blood glucose depleted appears to be fully oxidized but a large part also is represented by lactate accumulation (table 9.1, fig. 9.2). Assessing how different organs and tissues contribute to these overall changes in blood glucose and lactate should take us a long way toward understanding the outstanding breath-holding abilities of diving animals.

Impact of Diving on Brain Metabolism

There is good reason to suspect the brain as a key organ contributing to glucose depletion during diving, for the brain in most mammals displays an absolute dependence upon glucose as a carbon and energy source. So it is not surprising that glucose also is a suitable substrate for the seal brain. Recent studies show that glucose uptake rates by the brain are similar to rates in other mammals; during simulated diving, glucose consumption is elevated but only by about 140%, compared to the prediving state. Even if these estimates should contain small errors (due, for example, to admixture of venous blood from the brain and from other tissues such as skin and muscle), they are instructive, for they imply that brain metabolism in diving is not O_2-limited. If it were, glucose uptake would need to increase by up to eighteen-fold because of the energetic inefficiency of anaerobic glycolysis. Similarly, the fraction of glucose appearing as lactate would obviously rise. The estimated

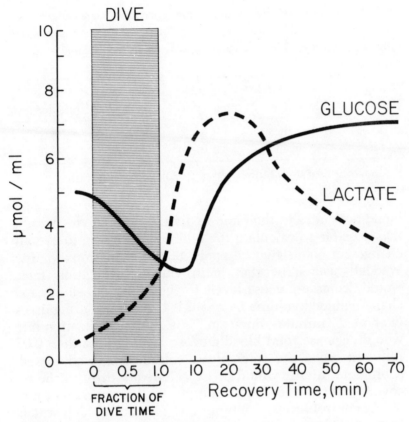

Figure 9.2 Generalized pattern of changes in glucose and lactate levels in whole blood during diving and recovery in the Weddell seal. (From Murphy et al. 1980, with modification.)

140% increase in glucose uptake, therefore, indicates that the brain's dependence upon anaerobic glycolysis does not rise greatly during simulated diving of up to 30 minutes' duration, despite PaO_2 levels as low as 25 mm Hg, which may be hypoxic to nondiving mammals. This is confirmed by lactate measurements showing that the lactate released by the brain accounts for 20 to 25% of the cerebral glucose uptake under both control and diving conditions (table 9.2). In the rat, the brain releases lactate because of limiting pyruvate dehydrogenase function, and this too would account for the data on the seal brain.

By assessing arteriovenous concentration gradients and

Table 9.2 Whole blood glucose and lactate concentration gradients across the brain of the Weddell seal prior to and during diving, expressed as the difference in μmol/ml between arterial and venous blood samples drawn simultaneously.

No. of samples	Condition	Lactate production	Glucose uptake	Glucose fermented (%)
20	Predive mean	0.14	0.28	25.0
12	Dive mean	0.16	0.28	20.0

Source: Data from Murphy et al. (1980), with modification.

blood flow rates to the brain in diving and prediving states, it is possible to calculate the brain contribution to overall changes in blood glucose pools. Such calculations for the Weddell seal show tbat brain glucose utilization rates would decrease glucose levels in the "central," slowly exchanging blood volume by about 0.4 μmol/ml during dives of about 20 minutes' duration. This glucose utilization rate would decrease total blood glucose level by less than 0.05 μmol/ml when the total blood volume was well mixed. (Complete mixing would be expected very early in the recovery process since cardiac output tends to overshoot prediving control levels, reaching values in the Weddell seal as high as 60 liters/min during the first minute of recovery.) The actual drop in blood glucose levels during such diving is about 1 μmol/ml (table 9.1). Therefore the brain can be eliminated from the list of tissues causing major changes in blood glucose levels during routine diving. As it turns out, the same is true for lactate.

Although in fermenting some glucose to lactate the seal brain is not unusual, it is exceptional in the amount (20 to 25%) that appears to be fermented, so the question arises as to whether the lactate production rate of the brain is high enough to significantly increase blood lactate levels during diving. The same kind of calculations as above demonstrate that brain anaerobic glycolysis would increase blood lactate concentration by only about 0.2 μmol/ml over a 20-minute diving period, assuming 8 liters of central circulating blood. This increase would be less than .03 μmol/ml when the blood was fully mixed, compared to an actual concentration change of about 2 to 3 μmol/ml. These calculations therefore clearly imply that during routine diving periods (of about 20 to 30 minutes' duration) brain metabolism on its own does not markedly alter blood pools of glucose or lactate. However, over many consecutive dives, the brain contribution to glucose and lactate concentration changes could become significant, so it is particularly interesting that changes mediated by brain metabolism are counteracted by metabolic processes in the lung.

Impact of Diving on Lung Metabolism

Since the lung was envisaged in earlier formulations of the diving response as a "favored" organ, it is of interest to consider how its metabolism is integrated with that of the brain and how it contributes to overall glucose and lactate pools in the blood. The picture emerging from arteriovenous concentration gradients across the lung is a mirror image of metabolism in the brain. That is, during diving the lung does not release lactate but rather utilizes it. At the same time, although glucose can be utilized by the lung, it can also be released at a low rate, a result consistent with the relative activities of glucose 6-phosphatase and hexokinase in the lung (table 9.3). There is currently no evidence suggesting *de novo* synthesis of glucose from lactate. Rather, the major fate of lactate taken up by the seal lung appears to be oxidation. This is indicated by in vivo experiments showing $^{14}CO_2$ is the only signficant metabolic derivative of injected ^{14}C-lactate on a single circulatory pass through the lung. In these experiments, a bolus of ^{14}C-lactate (about 4 μC/liter blood) plus dye and carrier lactate, rapidly injected into the right ventricle, served as a sensitive indicator of lactate oxidation during diving. Blood samples were taken every 20 seconds at two ports representing right (pulmonary artery) and left (aorta) sides of the circulation. Under these conditions, the majority of the $^{14}CO_2$ pulse appears on the left side of the heart, peaking shortly after injection; this first $^{14}CO_2$ peak could only have been generated by lung metabolism since at this time the lung is the only tissue (other than blood) to have received ^{14}C-lactate. For

Table 9.3 Enzyme activities in brain, heart, and lung of the Weddell seal expressed in terms of μmoles substrate converted/min/gm wet tissue weight at 37°C, pH 7.4, and saturating levels of substrates, cofactors, or coenzymes. Comparable values are included for the same organs in the ox.

Enzymes	Brain		Heart		Lung	
	Seal	Ox	Seal	Ox	Seal	Ox
Citrase synthase	17.8	16.8	28.8	61.7	1.52	6.6
Glutamate dehydrogenase	7.5	4.5	4.4	2.8	0.9	0.9
β-hydroxybutyrate dehydrogenase	0.4	0.3	2.4	2.8	1.7	0.2
β-hydroxybutyryl CoA dehydrogenase	3.4	2.2	16.0	21.6	1.2	5.2
Hexokinase	5.2	1.3	2.0	2.9	1.7	2.2
Glucose 6-phosphatase	0.6	—	0.5	—	0.7	—
Phosphofructokinase	8.6	8.6	16.7	14.0	3.7	4.7
Pyruvate kinase	167.3	196	217.5	133.1	45.6	98.0
Lactate dehydrogenase	229	128	1032	556	70	92

Source: Data from Murphy et al. (1980), with modification.

the duration of the test period (and presumably for the entire dive), lung oxidation of lactate exceeds lactate oxidation by other organs, a process which maintains consistently higher levels of $^{14}CO_2$ in the aortic blood than in the pulmonary artery (fig. 9.3).

Thus, in the seal lung, lactate can be used in preference to glucose as a carbon and energy source, a characteristic that subsequently has been shown in lung metabolism in two terrestrial species (rat and cow) as well as in the harbor seal. The functional significance of this arrangement during diving is that it spares glucose for other organs (particularly the brain) while minimizing the gradual accumulation of lactate in the central blood volume. In these functions, the lung may be assisted by the heart which may be particularly effective in reducing lactate accumulation in central blood during diving.

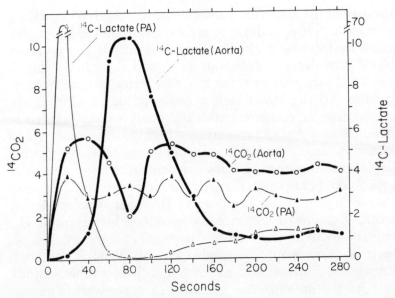

Figure 9.3 ^{14}C-U-lactate oxidation by Weddell seal lung in vivo after injection into the right ventricle. ^{14}CO$_2$ and labeled lactate were monitored at 20-second intervals in blood in the pulmonary artery and the descending aorta. The experiment was initiated after a stable bradycardia (heart rate of 15 beats/min) was established at 10 minutes into a 17-minute dive. (From Murphy et al. 1980.)

Impact of Diving on Heart Metabolism

Interestingly, just the opposite situation was initially envisaged for the heart (that is, lactate formation rather than utilization), which may in fact occur in emergency situations. There are several reasons for this view. In the first place, while the activities of hexokinase and other glycolytic enzymes are comparable to those in other species, heart lactate dehydrogenase activity and glycogen stores are the highest thus far measured in mammals (table 9.3). Moreover, glycogen is stored as large-diameter α-rosettes rather than the usual β-particles, a pattern typically observed only in organs storing unusual levels of glycogen. As in other species, heart LDH is potentially bifunctional, isozymes being present favoring either lactate formation or lactate utilization depending upon metabolic circumstances. Although all the above are consistent with a high potential for anaerobic glycogenolysis or glycolysis, it should be emphasized that during diving both cardiac output and coronary blood flow decrease by about 85%, and there is the rub. It can be shown that at these low flow rates, for the heart to produce all the blood lactate observed during diving, the concentration gradient across the heart would have to be over 20 μmol/ml, representing a highly unrealistic rate of lactate release. The same considerations apply for glucose depletion. Moreover, a strong dependence upon anaerobic glycolysis, because of energetic inefficiency, would necessitate a large increase in coronary flow per watt of cardiac work. If the heart were totally sustained by anaerobic glycolysis, about an eighteen-fold increase in coronary flow per watt of work would be needed. In the Weddell seal, it is known from measurements of blood flow, cardiac output, and arterial pressure that coronary flow per watt of cardiac work is essentially unchanged during diving. Thus, even if PaO_2 levels fall to levels that may be hypoxic in nondiving mammals, heart work during diving in the seal remains supported by oxidative metabolism. Whereas oxidative metabolism in the mammalian heart may be fired by a variety of substances (glucose, fatty acids, lactate), lactate is known to be preferentially utilized whenever concentrations rise above normal. That indeed is the situation developing through the diving period and it may explain why the Weddell seal heart retains exceptionally high levels of a lactate dehydrogenase kinetically suited for lactate oxidation.

Metabolic Model of the Central Organs during Diving

From the above data, the outlines of some major metabolic processes in the heart, lung, and brain of the seal can be discerned, allowing the construction of a simple, integrative model of events during diving. According to this model, the brain, lung, and heart display specific substrate preferences. The brain utilized glucose in a mixed aerobic–anaerobic metabolism that releases lactate during diving at a rate equivalent to nearly $\frac{1}{4}$ of the total glucose consumed. The lung and heart utilize lactate preferentially and thus diminish its accumulation in the central blood volume while simultaneously sparing blood glucose for the brain. Since cardiac work and perfusion are diminished, cardiac O_2 requirements must also be reduced by a similar factor (down to 15% of the nondiving state). The lung receives all the right cardiac output; so it is a fair assumption that its perfusion drops by at least a factor equivalent to the drop in cardiac output, and that its O_2 requirements too are reduced during diving. Thus, the metabolic organization of the brain, lung, and heart is designed to spare O_2 as well as glucose primarily for the brain while minimizing the accumulation of lactate.

The snag with this model is that it does not explain the magnitude of the observed changes in blood glucose and lactate levels during diving: in the first place, brain glucose utilization rate is not high enough to account for the large drop in blood glucose level that is observed, while with respect to lactate the situation is even further out of line. Not only is the brain production rate too low to account for observed changes, but lactate utilization by the heart and lung further diminishes its accumulation. So by elimination, the glucose depletion and lactate accumulation observed in the central blood volume must be largely due to metabolism of peripheral, hypoperfused organs and tissues.

Although the blood flow to many peripheral tissues is greatly reduced during diving, it is not zero. Skeletal muscle, liver, kidney, and skin, for example, together receive a significant fraction (about 15%) of cardiac output in the seal. In principle, an organism could compensate for reduced perfusion of any given organ during diving by storing O_2 (plus appropriate carbon substrate), and to some extent this must occur, since muscle myoglobin and glycogen levels in marine mammals are notably high. However, as already noted in Scholander's original study, (1940), myoglobin-bound O_2 stores are depleted in the early part of prolonged dives, and O_2 limitation would occur even more quickly in other hypoperfused tissues. Thus, it is widely accepted that during diving, many hypoperfused organs and tissues rely almost exclusively on anaerobic glycolysis. That is why there occurs a relatively close stoichiometry between blood glucose depletion and blood lactate accumulation (table 9.1). This interpretation also explains why glucose utilization and lactate production are so high: because markedly hypoperfused organs and tissues constitute the bulk of the animal, and because anaerobic glycolysis is energetically inefficient, requiring a large glucose consumption. Although in principle the above situation may prevail in most hypoperfused tissues during prolonged diving, there is a great deal of tissue specificity to metabolism and it is therefore informative to consider different tissues separately. Since muscle is such a large fraction of the whole animal and must remain active during diving (in prey capture, for example, or other activities), the metabolic consequences of diving in this tissue may have the largest effect on the organism as a whole. In this regard, only the dolphin has been examined in enough detail to illustrate the categories of adjustment coincident with the diving habit.

Metabolic Model of Dolphin Muscle

As discussed in chapter 5, glycolysis in nondiving animals is already a most effective anaerobic machine and it is further improved in its capacity and efficiency in dolphin muscle through only a modest number of modifications. Some of these influence glycolytic potential; some influence control. To increase potential, the steady-state concentrations of glycogen and of a few glycolytic enzymes are increased. Because muscle in marine mammals is hypoperfused during diving, it must rely less on blood glucose and more on muscle glycogen than would muscle in a terrestrial mammal; this is reflected in an increased phosphoglucomutase content. Similarly, an elevated aldolase activity contributes to tight regulation of the concentration of fructose 1,6-biphosphate, a key regulatory metabolite in the overall pathway. And finally, muscle cells in diving mammals must be prevented from becoming highly reduced and this requirement is reflected in high titers of M_4-type lactate dehydrogenase.

To retain control of the higher glycolytic capacity, the standard mechanisms discussed in chapter 5 undoubtedly come into play. In dolphin muscle, however, at least two additional modifications are known: muscle fructose biphosphatase activity in the dolphin is one of the highest thus far reported for any animal species, the enzyme appearing to function indirectly in glycolytic control; and muscle pyruvate kinase, although having a lower activity, occurs as a regulatory enzyme, sensitive to feedforward and feedback regulation.

In physiological terms, fructose biphosphatase catalyzes a reaction which is the reverse of that catalyzed by phosphofructokinase (PFK). Although phosphofructokinase is regulated by a large number of modulators, fructose biphosphatase seems to be under the regulation of only one effector compound: AMP. Adenosine monophosphate is a potent inhibitor of all fructose biphosphatases thus far examined, except for that in bumble bee flight muscle, where the enzyme may be involved in a thermogenic function. By designing both phosphofructokinase and fructose biphosphatase to respond (in opposite manner) to at least one signal in common (AMP), the sensitivity of the overall F6P $\leftrightarrow$ FBP conversion may be far greater than if only one of the enzymes were AMP-sensitive, or if only PFK were present.

In diving animals, there may be an added effect of temperature on fructose biphosphatase function. Over a decade ago, it was first observed that the AMP inhibition of fructose biphosphatase is unusually temperature-sensitive; the lower the temperature, the tighter the AMP binding, presumably because of stabilization of ionic interactions between the AMP pyrophosphate bridge and the fructose biphosphatase allosteric binding site. Recently, van Tol (1975) observed that in the absence of AMP, mammalian fructose biphosphatase shows an expected Q_{10} of 2, but that in the presence of physiological levels of AMP, the Q_{10} is 20. What it means is that a change of only 3°C can lead to a five-fold change in fructose biphosphatase activity. Diving animals are known to sustain a drop in body temperature of about this magnitude, and this would seem particularly important in peripheral organs which may go anoxic. The consequent reduction in energy metabolism would contribute to dropping temperatures and to a dramatic drop in fructose biphosphatase activity, thus facilitating F6P $\rightarrow$ FBP conversion catalyzed by PFK at a time when it is most needed for anaerobic glycolysis.

Unlike the enzyme in terrestrial mammals, muscle pyruvate kinase in diving vertebrates is under tight feedback inhibition by ATP, alanine, and probably citrate, and under strong feedforward activation by FBP, which sets the pyruvate kinase maximum potential in the range expected for highly active glycolysis. As with other regulatory pyruvate kinases, fructose biphosphate acts through two mechanisms: one directly activates the enzyme by increasing both enzyme–substrate affinity and the maximum velocity, while the other reverses the inhibitory effects of negative modulators. These regulatory characteristics are, in fact, commonly observed in muscle pyruvate kinases of lower animals (see chapter 3), but appear to have been lost in most mammals.

The reason diving animals have retained a tight control over muscle pyruvate kinase in part stems from a high reliance on glycolysis during diving. But more importantly, this enzyme fits the control requirements imposed on muscle at the end of the dive, that is, during anaerobic–aerobic transition, for inhibitory (alanine) control of pyruvate kinase at this time would contribute to dampening of glycolysis and thus to a sparing of glycogen for subsequent diving. The latter characteristic is a most telling consequence of the diving habit, for it reflects the need for swift, efficient transitions between anaerobic, carbohydrate-based fermentation and aerobic, fat-based oxidative metabolism. The control requirements imposed on muscle by this metabolic organization seem to call for unusually high levels of aspartate and alanine aminotransferases. The mitochondrial form of aspartate aminotransferase is designed to spark the Krebs cycle by increasing the availability of oxaloacetate at the same time that acetyl coenzyme A is being produced by β-oxidation. Alanine aminotransferase regenerates the α-ketoglutarate required for this process and leads to the accumulation of alanine. That is why alanine is such a good inhibitor of pyruvate kinase in dolphin muscle, for the greater the degree of Krebs-cycle activation, the greater the degree to which pyruvate kinase is blocked by alanine and the greater the degree by which carbohydrate is spared for future anaerobic excursions.

Metabolite Washout from Muscle during Recovery

The above summary model leads to some quite specific predictions on transient metabolite concentration changes in muscle during and following diving. Unfortunately, direct sampling of muscle has not been possible to date. On the positive side, however, because muscle is hypoperfused during diving, on reperfusion in recovery it is possible to see "pulses" of at least those metabolites that can diffuse out of the tissue into blood. Dominant among these is lactate, and it is satisfying that a hallmark feature of diving (lactate accumulation in hypoperfused tissues such as muscle and subsequent washout into the blood during recovery) would be predicted even if it had not already been repeatedly documented (fig. 9.2). A corollary of the lactate washout—that blood pyruvate levels should rise—also is now known.

Another expectation is that following diving alanine concentrations in muscle should increase while aspartate levels should drop. Unfortunately, no data are available on either amino acid. However, such an effect would explain the rise in whole-blood alanine levels following diving. With respect to alanine and amino nitrogen metabolism in general, it is relevant that glutamine is the chief form in which excess nitrogen is moved from muscle. In the rat, for example, three times as much amino nitrogen leaves muscle in the form of glutamine as leaves in the form of alanine. It is probable therefore that a large part of intracellular alanine, formed in part during transition to aerobic fat oxidation, is quickly deposited in a glutamine sink. Such a process would explain the observed rise in blood glutamine levels following diving in the Antarctic Weddell seal.

The picture of muscle metabolism that emerges thus implies the accumulation of metabolic end products during diving. Some fraction of this pool of metabolites leaks into the blood and leads to rising blood concentrations, but quantitatively this is a relatively modest process because of reduced perfusion. In recovery, when muscle perfusion is reestablished, larger pulses of these metabolites (particularly of lactate and glutamine) appear in the blood. What is their subsequent metabolic fate? To answer this question it is necessary to consider the contributions of different tissues to the clearance of lactate and glutamine from the blood.

Recovery Metabolism of the Brain, Lung, and Heart

The kind and amount of lactate dehydrogenase in the three central organs supply important clues as to their roles in lactate clearance during recovery. Interestingly, in all three organs in all diving species thus far studied, heart- and muscle-type subunits of lactate dehydrogenase are synthesized, and multiple isozymes occur in all three organs (fig. 9.4). All three organs have the potential for lactate utilization, catalyzed most effectively by H-type lactate dehydro-

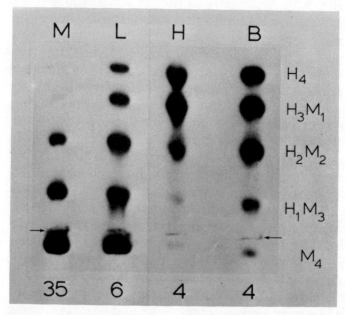

Figure 9.4 Electrophoretograms of lactate dehydrogenase isozymes in skeletal muscle (M), lung (L), heart (H), and brain (B) of the Weddell seal. The subunit composition of each tetramer isozyme is shown on the right. The numbers below refer to the ratio of pyruvate reductase activity/lactate oxidase activity at pH 7.4 and 37°C. (From Murphy et al. 1980, with modification.)

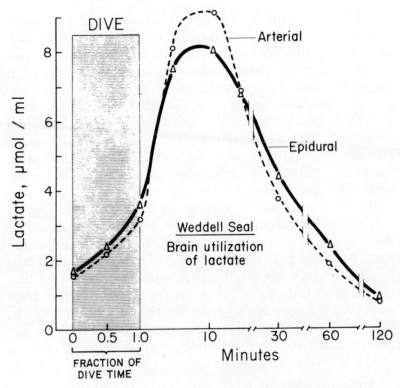

Figure 9.5 Arteriovenous lactate concentration gradient across the brain of the Weddell seal during diving and recovery. During the 20-minute dive, the brain was continuously producing lactate. During the large lactate washout occurring in recovery, the brain utilized lactate causing a reversal of the arteriovenous gradient. (From Murphy et al. 1980.)

genases. Empirically this is indicated by the ratio of pyruvate reductase activity to lactate oxidase activity, which is much lower than for skeletal muscle lactate dehydrogenases. It is not surprising, therefore, that during recovery, when lactate washout peaks are high, the seal brain switches from lactate release to lactate uptake. The critical arterial blood concentration range for brain lactate uptake in the seal appears to be about 6 μmol/ml (fig. 9.5); when this is surpassed, either in diving recovery or by lactate infusion, the brain vigorously consumes lactate, generating an AV gradient of over 1 μmol/ml. If all the lactate consumed was fully oxidized, it could support a metabolic rate of 9

μmol ATP/gm/min, assuming that blood flow to the brain was normal. This value is equal to or greater than that sustainable by glucose metabolism and indicates that lactate metabolism under these conditions can readily supply all of the energy demands of the brain.

Similar data are not yet available for the heart. However, qualitative lactate dehydrogenase isozyme patterns for the brain and heart are similar, while overall activities are five-fold higher in the heart. Thus conditions are suitable during the lactate washout for blood lactate to serve as a carbon and energy source for the seal myocardium, as occurs under conditions of high blood lactate levels in other mammalian species as well. A similar metabolic capacity for lactate utilization is displayed by the lung, particularly at the high blood-lactate levels found during the recovery process. It is therefore a fair conclusion that all three central organs contribute to clearing lactate from the blood during recovery from diving (figs. 9.6 and 9.7). In this process they are undoubtedly aided and, indeed, because of size and perfusion rates, probably overshadowed, by the liver and kidney.

Recovery Metabolism of the Liver and Kidney

The reason we have a fair appreciation of metabolic events that must occur in the liver following diving is that a lot of blood metabolite information is available that bears on this matter. On recovery, the reestablishment of liver perfusion and elevating PO_2 levels are probably the first events heralding the return to normal oxidative metabolism. In metabolic terms, this situation should lead to energy-saturated cells quite quickly. At the same time, lactate is being washed out of peripheral organs, particularly muscle, and blood lactate levels can rise dramatically. An important fate of blood lactate is reconversion to glucose. Elevated levels of glucagon, known to occur early in recovery, favor liver gluconeogenesis and the flow of lactate carbon back into glucose.

Early in recovery, catecholamines, and norepinephrine in particular, also are known to be elevated in the blood, which is one reason why the adrenal gland remains rather well perfused during diving. Of several metabolic consequences, a critical one in the liver is epinephrine activation of adenyl cyclase and the subsequent activation of the glycogen-mobilizing enzyme cascade. This process would favor the release of glucose from glycogen.

These two events by themselves are capable of increasing glucose release from the liver, which in turn should be evident in rising blood glucose levels during recovery. Since blood glutamine levels are also elevated in the postdiving recovery period, glutamine uptake by the liver, in addition to serving as a precursor to urea, would augment liver gluconeogenesis. For all these reasons, liver production of glucose should rise and it is not surprising that blood glucose levels quickly return to normal during recovery from diving. Indeed, shortly after diving, blood glucose levels may be higher than in the prediving state, implying that these control mechanisms actually overshoot starting conditions. A similar situation occurs in kidney metabolism during diving–recovery cycles.

In physiological terms, the striking feature of the kidney in diving animals of course is that it is treated as a peripheral organ and is vasoconstricted during diving. In the harbor seal, the kidney can be fully isolated from circulation for at least an hour with no irreversible detrimental effects, and with essentially instantaneous recovery. Although such drastic circulatory isolation does not occur in vivo, the seal kidney undoubtedly possesses a most capable anaerobic metabolism charged entirely by endogenous substrate. But whether or not it is utilized in all parts of the nephron structure is not known. Because in seals during routine diving PO_2 seldom falls below about 25 mm Hg and because of the occurrence of numerous, giant mitochondria in the proximal tubules, it is tempting to believe that at least part of kidney metabolism remains aerobic during diving.

During recovery from diving, the metabolic situation is somewhat clearer, for at least the following information is available: pH is low, lactate and glutamine levels are high, while other parameters (PO_2, PCO_2, glucose levels, fatty acid levels) are probably normal or nearly so, either because they are not expected to change during diving (for example, fatty acids) or because their return to normal is rapid compared to the lactate and glutamine recovery patterns (for example, PO_2). Since glutamine is used preferentially under acidic conditions in nondiving mammals, it is probable that glutamine initially contributes the largest fraction to overall energy metabolism, with lactate as the second most important carbon source. As the pH is stabilized, the impor-

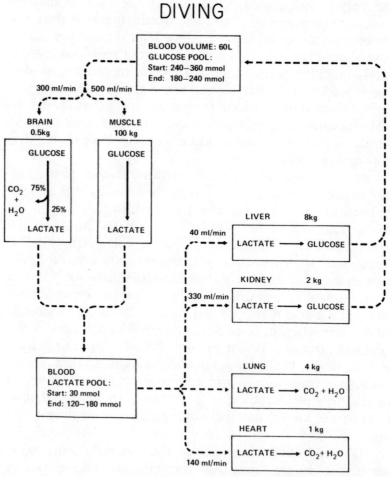

Figure 9.6 Summary of metabolic and physiological events known to occur in the Weddell seal during diving. (After Murphy et al. 1980 and Zapol et al. 1979.)

RECOVERY

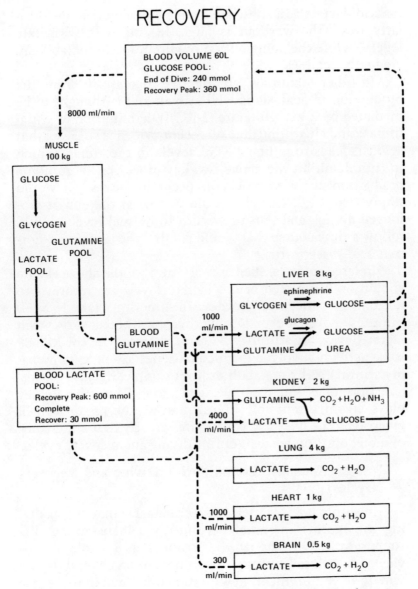

Figure 9.7 Summary of metabolic and physiological events known to occur in the Weddell seal during recovery from diving. (After Murphy et al. 1980 and Zapol et al. 1979.)

tance of lactate as a substrate source probably rises (particularly when the washout is large, leading to high lactate levels), while the contribution of glutamine to metabolism gradually declines.

In other mammals, an adaptive increase in ammonia production in acidosis is well known and is thought to be mediated by 2-ketoglutarate (2-KGA) deinhibition of glutaminase and gluatmine uptake: recent evidence suggests that acidosis leads to falling 2-KGA levels in the kidney which in turn deinhibit the above two key processes in ammonia production. If the same events occur in the seal, it would imply that 2-KGA levels should be low at the end of prolonged diving, and that the return to normal levels should follow a time course that would parallel the return to a normal acid–base status.

In large animals, such as seals, most of the above events should be completed before lactate levels are returned to normal since the latter process can sometimes take 15 to 30 minutes or even longer. During this phase of recovery, when lactate levels are still high, we would anticipate the kidney to be actively involved in gluconeogenesis, a process probably stimulated by metabolites (for example, free fatty acids) and hormones (for example, glucagon). Thus the kidney plays an important role in augmenting the production of glucose by the liver and contributes greatly to maintenance of metabolic homeostasis during diving and recovery cycles.

Major Metabolic Events during Diving and Recovery Cycles

In summary, it is useful to reemphasize that during diving the marine mammal becomes a self-sustaining life-support system. Although the organism as a whole may be viewed as a closed system, its component organs and tissues remain in communication with each other and in this sense are open systems. Although communication channels remain open, the quantitative metabolic exchanges between tissues and organs are strikingly modified by circulatory adjustments during diving in order to extend the breath-hold period. The primary carbon and energy source sustaining most tissues and organs during diving is blood glucose. Its metabolism at this time can be viewed as occurring in three phases (glucose $\rightarrow$ lactate, lactate $\rightarrow$ CO_2 + H_2O, and lactate $\rightarrow$ glucose) with each phase occurring in different tissues or organs (fig. 9.6). Phase I, anaerobic glycolysis, is energetically inefficient, requires a large consumption of glucose, and occurs primarily in hypoperfused tissues and organs such as muscle; but even in the brain, about $\frac{1}{5}$ to $\frac{1}{4}$ of the glucose taken up is represented by lactate release. Rising blood lactate levels during diving (due in part to release from the brain but mainly to a slow drainage from hypoperfused organs and tissues, mainly muscle) set the stage for phase II: complete oxidation of lactate elsewhere in the body, primarily in the heart and lung. This process is functionally advantageous because it reduces lactate accumulation in central circulating blood and spares glucose for the brain. These cooperative metabolic interactions between brain, lung, and heart require only modest enzyme modifications and are probably quite critical because the central organs are sustained during diving by a slowly exchanging, reduced blood volume (in the seal, this is about $\frac{1}{10}$ to $\frac{1}{5}$ the total blood volume).

In phase III the fate of lactate is determined by still other organs, the kidneys and liver. Although perfusion of both is reduced, it is a long way from zero; the kidney, for example, receives 3% of cardiac output during diving and absolute blood flow rates are greater than to the heart. Lactate-primed gluconeogenesis in the kidneys and liver at this time, even if occurring at a slow rate, could slow down the rate of blood glucose depletion. This process may explain why in the Weddell seal during unusually long dives the rate of blood glucose depletion gradually slows down and an equilibrium point appears to be reached; at this stage, glucose uptake from the blood is matched by glucose (production in the liver and kidneys, then) entry into the blood. In this view, the purpose of circulatory adjustments is not only to conserve oxygen for the central organs (as originally suggested) but also to allow the development of crucial metabolite exchanges among different tissues and organs that sustain metabolic homeostasis for extended breath-hold diving periods.

In recovery (fig. 9.7), when normal perfusion patterns are reestablished, the metabolic situation is less novel and is dominated by a large washout of lactate and glutamine from hypoperfused tissues such as muscle. All three central organs appear capable of contributing to the clearance of blood lactate, but as in nondiving mammals, the liver and kidneys may make even larger contributions to this function. In the central organs the fate of lactate is probably mainly oxidation, while in the liver and kidneys it is probably gluconeogenesis, primed by elevated glucagon levels. High epinephrine levels would also contribute to mobilizing liver glycogen to glucose at this time and thus to recharging blood glucose reserves. The kidneys and liver are also responsible for the clearance of a pulse of blood glutamine. In the kidney this process contributes to reestablishing acid–base balance, while in the liver it sets the stage for incorporating waste nitrogen into the urea sink while most of the glutamine carbon probably is incorporated into glucose, in this way augmenting the lactate-primed reestablishment of glucose homeostasis.

Suggested Readings

ANDERSEN, H. T. 1966. Physiological adaptation in diving vertebrates. *Physiol. Rev.* 46:212–243.

ELSNER, R., FRANKLIN, D. L., CITTERS, R. L., and KENNEY, D. W. 1966. Cardiovascular defence against asphyxia. *Science* 153:941–949.

HOCHACHKA, P. W. LIGGINS, G. C., QVIST, J., SCHNEIDER, R., SNIDER, M. Y., WONDERS, T. R., and ZAPOL, W. M. 1977. Pulmonary metabolism during diving: conditioning blood for the brain. *Science* 198:831–834.

HOCHACHKA, P. W., and MURPHY, B. 1979. Metabolic status during diving and recovery in marine mammals. In Robertshaw, D., ed., *Intl. Review Physiology*, vol. 3., *Environmental Physiology*. Baltimore: Univ. Park Press. 253–287.

HOCHACHKA, P. W., and STOREY, K. B. 1975. Metabolic consequences of diving in animals and man. *Science* 187:613–621.

MURPHY, B., ZAPOL, W. M., and HOCHACHKA, P. W. 1980. Metabolic activities of the heart, lung, and brain during diving and recovery in the Weddell seal. *J. Appl. Physiol.*, 48.

NEELY, J. R., and MORGAN, H. E. 1974. Relationship between carbohydrate and lipid metabolism and the energy balance of heart muscle. *Annu. Rev. Physiol.* 36:413–459.

ROBIN, E. 1979. Glucoregulation in aquatic mammals as a model for the defense of brain glucose requirements. Personal communication.

SCHOLANDER, P. F. 1940. Experimental observations of the respiratory function in diving mammals and birds. *Hvalradets Skrifter, Norske Videnskaps Acad. Oslo* 22:1–131.

———. 1962. Physiological consequences of diving in animals and man. *Harvey Lectures* 57:93–110.

SIESJO, B. K., and NORDSTROM, C. H. 1977. Brain metabolism in relation to oxygen supply. In Jobsis, F. F., ed., *Oxygen and Physiological Function.* Dallas: Professional Information Library. Pp. 456–475.

STOFF, J. S., EPSTEIN, F. H., NAIRNS, R., and RELMAN, A. S. 1976. Recent advances in renal tubular biochemistry. *Annu. Rev. Physiol.* 38:46–68.

TIERNEY, D. F. 1974. Lung metabolism and biochemistry. *Annu. Rev. Physiol.* 36:209–231.

VAN TOL, A. 1975. On the occurrence of a temperature coefficient (Q_{10}) of 18 and a discontinuous Arrhenius plot for homogeneous rabbit muscle fructose diphosphatase. *Biochem. Biophys. Res. Commun.* 62:750–756.

WOLFE, R. R., HOCHACHKA, P. W., TRELSTAD, R. L., and BURKE, J. F. 1979. Lactate oxidation in perfused rat lung. *Am. J. Physiol.* 236:E272–E282.

ZAPOL, W. M., LIGGINS, G. C., SCHNEIDER, R. C., QVIST, J., SNIDER, M. T., CREASY, R. K., and HOCHACHKA, P. W. 1979. Regional blood flow during simulated diving in the conscious Weddell seal. *J. Appl. Physiol.* 47.

Epilogue

There are two sides to the coin we have been examining in our exploration of animal anaerobic mechanisms in the preceding chapters. One side, favored by classical biology, arouses our interests in the spectacular diversity of life forms and their impressive variety of ways of dealing with limited oxygen availability. The other side, favored by classical metabolic biochemistry and physiology, arouses our interests in unifying themes in hypoxia and anoxia tolerance, themes that are shared by all life forms, by seals, squids, bivalves, and birds, by *Nautilus*, *Protopterus*, marine mammals, and man. A major goal of this book has been to fuse these two extreme points of view: to survey and use natural variations in hypoxia tolerance for gaining a better understanding of common principles and problems in anaerobiosis. The best way of judging the success of such a comparative endeavor is to consider the insights that arise from it. These turn out to be unifying and surprisingly simple.

When an organism encounters conditions of limiting oxygen availability, for whatever internal or external reasons, it must in effect become a self-sustaining life-support system until it returns to more favorable conditions. In the design of such life-support systems, provisions must be made at the level of each cell for ample fermentable substrate, for redox maintenance in cytosol and mitochondria, and for energy generation in the form of ATP. These are closed-system provisions and each is requisite by any cell, tissue, or organ considered in isolation; for such reason, the metabolic machinery assuring these provisions is usually considered to be highly conservative. Analysis of a broad spectrum of most capable animal anaerobes indicates, in contrast, that all these primary requisites for living without oxygen are in fact readily adjustable and supply the raw material for adaptive modifications.

Change can occur in many parts of the closed system. Thus, the amount of glycogen stored and the storage form (as β-granules, rosettes, glycogen bodies, or glycogen seas) can be varied according to the need for efficiency of packing. Glycogen as a storage form of energy can be supplemented by a large pool of free amino acids (for example, aspartate) for simultaneous fermentation with glycogen. Overall metabolic organization in such cases may be modified in order to integrate the processes of glycogen and amino acid fermentation. Often this leads to multiple mechanisms for balancing redox potentials and the concommitant generation of multiple anaerobic end products. Almost invariably, such metabolic reorganization also leads to increased ATP yield and the generation of a greater diversity of coupling intermediates necessary for sustaining anabolic functions. Finally, not only are the kinds of enzymes, and hence the available metabolic pathways, regulated, so also are enzyme amounts and enzyme kinetic properties, which supply additional ways of expanding the cell's anaerobic potential and anoxia tolerance.

To improve anoxia tolerance, it is therefore already clear that a great deal of adjustment can and does occur at the cellular (closed-system) level. Although numerous additional mechanisms will surely be uncovered by future work, they plus those already described constitute only one part of a two-part strategy that organisms utilize in dealing with oxygen lack. The second-level strategy depends upon the fact that most organs and tissues do not become wholly closed systems during hypoxia or even during anoxia. Rather, via the blood vascular system, they remain in close communication with each other. In this sense, organs and tissues in hypoxia are open-systems and retain the possibility of functionally useful communication channels. The second-level strategy of hypoxia adaptation depends upon the establishment of key metabolic interactions or exchanges between tissues and organs that contribute to extending the hypoxia tolerance of the organism as a whole.

The possible number of such metabolic exchanges is theoretically very large, but only a few simple ones are thus far identified. In some organisms (hypoxia-adapted fishes, such as the goldfish, for example), it appears that glycogen remains the primary storage substrate for fermentation, but lactate does not accumulate to very high levels. This is because lactate formed by anaerobic glycolysis is further metabolized at sites remote from sites of production, with the concommitant formation of relatively harmless additional end products which may be readily released into the outside water. Although wasteful of carbon (a two-carbon compound, probably ethanol, is washed out into the external medium), it contributes to extending the anoxia tolerance of the whole organism by preventing a large accumulation of lactate; some such mechanism might indeed have been anticipated sinces fishes are unable to maintain large bicarbonate-based buffering reserves. Similar metabolic interactions between the brain, heart, and lung during diving minimize the accumulation of lactate in the central blood and thus may also be involved in extending the period of breath-hold diving in marine mammals.

Such metabolic interactions between tissues and organs may differ in nature. Often metabolite exchanges may be in the form of substrate-product cycling; for example, lactate formed by the brain and peripheral organs during diving may be utilized as a substrate by the heart and lung in preference to blood glucose. Or, metabolic exchanges between tissues and organs in differing degrees of hypoxia may involve hydrogen cycling (that is, exchange of metabolites in different oxidation states). The physiological purpose of lactate cycling is to conserve glucose for the brain and to minimize lactate accumulation in the central blood. The physiological purpose of hydrogen cycling is to retain redox balance in the most hypoxic tissues while conserving oxygen for the brain or other highly oxygen-dependent tissues. Thus, as at the closed-system level, it is already clear that a lot of adjustment can and does occur at the open-system level involving metabolic interactions between tissues and organs. While clarifying the meaning and importance of these second-level mechanisms in hypoxia tolerance remains an important job for the future, their obvious usefulness is well illustrated by their widespread occurrence in phylogenetically very diverse groups.

Finally, it will be evident that there must occur a critical interplay between closed-system and open-system modifications. Change and adjustment at a cellular level will be reflected in organ and tissue level metabolic capabilities, and hence will determine what metabolic interchanges are possible or impossible. To use the anoxic goldfish for illustrative purposes again, it is evident that the further metabolism of lactate would best be localized to tissues and organs with heart-type lactate dehydrogenases which are kinetically suited for lactate oxidase (and hence lactate scavenging) function. Not surprisingly, it is in such a tissue (red muscle) that the process of anoxic lactate oxidation proceeds most vigorously and that the activity of ethanol dehydrogenase, catalyzing the formation of ethanol as an ultimate anaerobic end product, is the highest. Although fascinating in its own right, the importance of the goldfish model stems from the principle it illuminates; namely, at the open-system level, achieving metabolic integration of different organs and tissues in anoxia requires, at the cellular level, that the synthesis and maintenance of key metabolic enzymes (lactate and ethanol dehydrogenases, in this case) be closely coordinated with each other. Indeed, this kind of coordination completes the complex control networks required between open and closed system levels. By and large, however, metabolic signals and mechanisms allowing such coordination are not known, and it is the deciphering of design rules underlying integrative functions of tissues and organs in anoxia or hypoxia that must remain as a major challenge for future research.

Index

Abyssal fish, 101–104
Acetate, 15, 21, 27, 108–109
Acetate thiokinase, 108
Acetyl coenzyme A, 7, 24–25, 40, 107–109
Acidosis, 166
Adaptation: of equilibrium enzymes, 65–67; of PFK, 74; air breathing, 117; of anaerobic metabolism, 1–17
Adenosine monophosphate (AMP), 70, 76, 157
Adenosine triphosphate, see ATP
Adenylate, 9–12, 35
Adenylate coupling, 9–10
Adrenal gland, 162–163
Aerobic/anaerobic potential, 103, 131, 137, 154–156
Aerobic glycolysis: integration with anaerobic, 79–99, 103, 131
Affinity: of enzyme-substrate interaction, 63, 65
African lungfish, see Protopterus
Air-breathing fish, 117–143
Alanine, 32–33, 34, 37–38, 112–113, 158–159; in bivalves, 27, 29, 42, 54–55
Alanopine, 27, 33–39, 54–55
Alanopine dehydrogenase (ADH), 34, 35–36, 37–39

Aldolase, 66–68, 156
α-particles (glycogen), 3, 127
Amino acid metabolism: coupled to glucose metabolism, 21–24, 27–40, 42–58; coupled to lipid formation, 112
Ammonia, 112, 166
Anaerobes, 2. See also Helminths
Anaerobic/aerobic potential, 103, 131, 137, 154–156
Anaerobic/aerobic transition, 158
Anaerobic end products, 7, 8, 77–78; in helminths, 15–16; in bivalves, 27–28; in oyster heart, 30–39; in goldfish and carp, 105–114
Anaerobic glycolysis, 1; and coupling intermediates, 10–12; in cephalopods, 42, 57–58; and dehydrogenase systems, 54; key elements of, 60–78; and redox balance, 77–78; integration with aerobic, 79–99, 103, 131; in tuna white muscle, 85–95; and brain enzymes, 121–124; in diving marine mammals, 145; in seal, 150–151, 155–156
Anaerobic metabolism, 1–13; carbon dioxide fixation in, 16–19; energy yield, 25, 40, 57–58
Anaerobiosis, 60

Anoxia, 9–10, 57, 122; time course, 12, 27; tolerance, 60–62, 171–173. *See also* Hypoxia
Antarctic Weddell seal, *see* Seal
Apnea, 146
Arapaima: brain enzymes, 122–124; white and red muscle, 124–127; heart, 136; kidney and gills, 137–141; liver, 141
Arginine: coupled to glucose metabolism in cephalopods, 42–58
Aruana, 137–138, 141
Ascaris, 16, 18
Aspartate, 30, 37, 79–80, 159
Aspartate aminotransferase, 36, 79–80, 113, 134, 158–159
Atkinson, D. E., 9
ATP, 1, 25, 60, 122, 137; sources during anoxia, 9–10; in control of PFK, 73–76; in tuna white muscle, 87–88
Avian sperm, 81–82

Bacteria, 102
Baikal, Lake, 103
Bee flight muscle, 83–85, 87
β-particle (glycogen), 2, 127
Binding, 62–65
Bivalve molluscs, 6, 27–40
Blazka, P., 106
Blood flow, 118–121, 146–148, 154, 155
Blow fly flight muscle, 81
Bradycardia, 104, 140
Brain, fish, 120, 122–124
Brain, mammalian, 74–76, 94–95, 122; impact of diving on, 148–151, 167; impact of recovery on, 160–162, 167
Brain enzyme adjustments, 121–124
Brain lactate uptake, 160–162
Burrowing fish, 131, 133

Calcium, 70
Calcium carbonate, 28

Capillary density: lungfish muscle, 127, 129; tuna muscle, 127, 129–131
Carbohydrate, 107, 148, 159; coupled to amino acid metabolism, 24, 38–39
Carbon, 62, 66
Carbon dioxide: as anaerobic end product, 106–108, 114; fixation by helminths, 15–25; fixation by bivalves, 28–32
Carbon dioxide/bicarbonate system, 19
Carbon monoxide, 107
N-Carboxyethylalanine, *see* Alanopine
Cardiac output, 104, 119–121, 146–148, 154, 155
Carp, 101, 105
Cascade control of enzymes, 70–73
Catecholamines, 162
Cephalopods, 6, 42–58
Cestodes, 16
Citrate synthase, 122–123, 126, 134, 137
CNS, 122. *See also* Brain
Control of enzymes: in glycolysis, 70–77, 79–97. *See also specific enzymes*
Coulter, G. W., 103
Coupling intermediates, 10–12, 25
Crassostrea, 27
Creatine phosphate, 75, 89, 91–92
Cuttlefish, 50, 51–53
Cytosol redox regulation, 6–7; in bivalves, 37–39; in cephalopods, 42, 55–56; and hydrogen shuttle, 79–83

Dehydrogenase: types of, 6–7. *See also* Glutamate dehydrogenase; Malate dehydrogenase; *etc.*
Dehydrogenase doubles: in anaerobic systems, 53–56; in mixed aerobic/anaerobic systems, 85–98

Dihydroxyacetone phosphate (DHAP), 69, 83–85
Diving: and recovery, glucose metabolism during, 141–143; impact on brain, 148–151, 167; impact on lung, 151–153, 167; impact on heart, 153–154, 167; metabolic model of central organs during, 154–156; major metabolic events in, 166–167
Diving fish, 117–143
Diving marine mammals, 145–168
Diving response, 145–146
Dolphin muscle, 156–159

Electron transfer system (ETS), 79
Endoplasmic reticulum (ER), 3
End products, *see* Anaerobic end products
Energy of binding, 62–63
Energy yield, 25, 40, 57–58, 109
Enzyme levels: in helminths, 18–19; in oyster heart, 30; in cephalopods, 48, 51, 53, 55; in tuna tissues, 86; in air-breathing fish, 123, 125, 135, 140; in seals, 152
Enzymes: regulatory versus nonregulatory, 62–65; criteria of adaptation, 65–67; in control of glycolysis, 67–77; adjustments in brain, 121–124; in fish hearts, 134. *See also specific enzymes*
Epinephrine, 70, 163, 168
Equilibrium constant for glycolysis, 10
Equilibrium enzymes, 62, 65–67
Erythrocytes, 75
Estivation, 131, 133
Ethanol, 108–109
Ethanol dehydrogenase, 108–109
Exercise: effect on glycogen deposits, 4–5

Fairbairn, D., 16
Fasciola, 16, 21, 42

Fat, 131. *See also* Lipids
Fatty acids, 21–24, 81–82, 109–112
Fersht, A., 63, 65, 69
Fish: hypoxia-adapted, integrative mechanisms in, 100–115; air-breathing, 117–143
Flight muscle: insect, 80, 81, 83–85, 87
Food: in abyssal waters, 103
Fructose biphosphatase (FBPase), 141, 156–158
Fructose biphosphate (FBP), 28, 77, 141
Fructose 6-phosphate (F6P), 73–74
Fumarate reductase, 24, 42
Futile cycles, *see* Substrate cycling

Gastrointestinal parasites, *see* Helminths
Gills, 137–141
Glucagon, 70
Glucocorticoids, 70
Gluconeogenesis, 53, 141
Glucose, 1, 16, 32–33, 70–72; coupled to amino acid metabolism, 21–24, 27–40, 42–58; test of anaerobic functions, 98–99; in goldfish, 105, 107; in air-breathing fish, 121; during diving and recovery, 141–143, 163–168
Glucose 6-phosphate (G6P), 72
Glutamate, 79–80
Glutamate dehydrogenase (GDH), 6–7, 44, 126, 134
Glutamate-oxaloacetate transaminase, 125–126, 134
Glutamine, 159, 160–166
Glyceraldehyde 3-phosphate (GAP), 68–69
Glyceraldehyde 3-phosphate dehydrogenase (GAPDH), 1, 37–39, 42, 60
α-Glycerophosphate (α-GP), 77; cycle, 80–81, 82, 83–85, 87–88, 93

α-Glycerophosphate dehydrogenase (α-GPDH), 6, 54, 77–78, 83–85, 87–95
Glycogen, 1, 60; in goldfish, 105, 107; in fish heart, 133–134; in seal, 155
Glycogenolysis, 72
Glycogen phosphorylase, 62, 70–73
Glycogen storage, 2–5; improving, 60–62; in lungfish, 127–131, 133
Glycogen synthetase, 70–72
Glycolysis, see Aerobicglycolysis; Anaerobic glycolysis
Glycolytic potential: adjustable components, 1–13; in dolphin muscle, 156–158
Goldfish, 101, 104, 105–115, 173

Haldane, J. B. S., 62
Heart, fish, 120, 121, 132–137
Heart, mammalian, 67–70, 75–76, 95–96; hydrogen shuttle in, 80–81; impact of diving on, 153–154, 167; post-diving effects on, 160–162, 167
Helminths, 15–25
Hexokinase, 62
Hydrogen shuttles, 79–80, 126, 134
3-Hydroxybutyrate dehydrogenase, 7
Hymenolepsis, 16, 18–19
Hypoxia, 104–105, 110–112, 121–124, 171–173. See also Anoxia
Hypoxia adaptations, 24–25, 39–40, 57–58
Hypoxia-adapted fish: integrative mechanisms in, 100–115

Insect flight muscle, 80, 81, 83–85, 87
Insulin, 70
Intestinal parasites, see Helminths

Isovalerate, 21–22, 23, 42
Isozymes: LDH, 50–51, 53, 66; ODH, 50–53; PFK, 75

Janaasch, H. W., 102

k_{cat}, 63; relationship to K_m, 63–69
K_m, 63; relationship to k_{cat}, 63–69; relationship to [S], 65–69. See also specific enzymes
Kesbeke, G., 107
Keto acids, 34
Keto carboxylate dehydrogenase reaction, 24
2-Ketoglutarate, 79–80, 166
Kidney: fish, 121, 137–141; mammalian, 138–141, 162–168

Lactate, 60, 66, 77; in helminths, 15, 19; in bivalves, 27; in cephalopods, 53; in goldfish, 105–106, 107, 113; in air-breathing fish, 121, 136; in Arapaima, 126–127, 139, 141; during diving and recovery, 141–143, 159, 160–168; in seal, 148–156
Lactate dehydrogenase (LDH), 6–7, 60, 77–78; in tuna muscle, 3, 61, 86, 87, 89–94, 97–99; in helminths, 15; in cephalopods, 30, 42; compared to ADH, 34, 35, 37; compared to ODH, 43–44, 53–54; isozymes, 50–51, 53, 66; in hydrogen shuttle, 82–83; in Arapaima, 122, 139, 141; in lungfish and Synbranchus, 134; bifunctional in heart, 134–136; in seal, 154; in dolphin, 156
Lepidosiren, 5, 119, 120, 127, 131. See also Lungfish
Leucine, 21–22, 23, 42
Lipids, 109–112, 131, 133–134
Liver, 67–70, 74–76, 81, 94–95; during diving and recovery, 162–168

Lung: during diving and recovery, 151–153, 160–162, 167
Lung lactate oxidation, 151–153
Lungfish, 106, 118–121, 127–132; and *Synbranchus* heart, 132–137

Malate, 18, 113
Malate-aspartate shuttle, 79–80, 82–83, 95, 134
Malate dehydrogenase (MDH), 6, 33, 42, 35–36, 113, 134; interaction with ADH, 37–39, 54–55; interaction with ODH, 55–56; in malate-aspartate shuttle, 79–80, 82–83; tissue control, 95–99; in *Arapaima* muscle, 124–126
Mammalian sperm, 81–82
Mammalian tissue: hydrogen shuttles in, 79–80
Mammals, 110, 112; diving marine, 145–168
Mathur, G. B., 101
Metabolic model: during diving, 154–159, 166–167
Metabolic regulation: elements of, 60–77
Microbes, 102
Mitochondria: in *Arapaima*, 124, 125–126; in lungfish, 127, 129, 131, 133
Mitochondrial redox regulation, 7–9, 40, 55; and hydrogen shuttle, 79–83
Muscle: glycogen bodies associated with, 3–4; and control of glycolysis, 61–62, 67–77; role of equilibrium enzymes in, 65–67; *Arapaima*, 124–127; lungfish, 127–132; dolphin, 156–159; during recovery, 159–160. *See also* Tuna red muscle; Tuna white muscle
Mussel, 27
Myoglobin, 155
Myotome, 124, 126

Mytilus, 27

NADH, 6, 7, 35, 44, 47, 66, 91–95; and double dehydrogenase systems, 54–57, 85–98; in hydrogen transfer, 83–85. *See also* Redox balance
Nautilus spadix muscle: ODH in, 48–49, 50; MDH in, 55–56, 97
Nematodes, 16
Nephron, 138. *See also* Kidney
Norepinephrine, 162

Octopine, 43, 45, 50, 51–53
Octopine dehydrogenase (ODH), 6, 7, 42–56; compared to ADH, 34; compared to LDH, 43–44, 53–54; isozymes, 50–53; interaction with MDH, 55–56
Octopus mantle muscle: ODH in, 44–49
Oxaloacetate, 36
Oxidation-reduction balance, *see* Redox balance
Oxidative metabolism, role of LDH in, 134–137
2-Oxybutyrate, 34
Oxygen uptake: in fish, 101–103, 120, 121–124, 126, 137; in mammals, 120, 146–148
Oyster, 27, 97
Oyster heart, 30–39, 54–55

Parasites, *see* Helminths
pH: in helminths, 19; in bivalves, 27, 36; in glycolytic control, 76–77, 89, 97; in hypoxia in goldfish, 113; in diving metabolism, 163
Phosphoenolpyruvate (PEP), 16–19, 28
Phosphoenolpyruvate carboxykinase (PEPCK), 18–19, 30
Phosphofructokinase (PFK), 62, 73–77, 122, 139, 141, 157
Phosphoglucomutase, 156
Phosphoglycerate kinase, 1, 60

Pressure: effect on oxygen uptake, 102
Proline shuttle, 81–82
Propionate, 15, 21–23, 27, 29
Prosser, C. L., 105
Protopterus, 119, 120, 131. *See also* Lungfish
Pseudemys, 122
Pyruvate, 33, 34, 44–49, 66, 77, 107, 159
Pyruvate dehyrogenase, 107
Pyruvate kinase (PK), 1, 60, 62; in helminths, 16–19; variants, 64–65, PFK integration with, 76–77; in *Arapaima*, 122, 139; during diving, 157–158
Pyruvate-lactate shuttle, 81–82

Q_{10}: in tuna muscle, 89; for O_2 uptake in fish, 102; unusual value for FBPase, 157

Rabbit heart, 109
Rabbit muscle, 92–94
Rangia, 27
Rasbora, 101
Rat heart, 109
Recovery metabolism post-diving, 159–168
Red muscle: *Arapaima*, 124–127; lungfish, 127–132. *See also* Tuna red muscle
Redox balance, 1, 6–9; in helminths, 24; in bivalves, 31–33, 37–40; in cephalopods, 42, 54–56; maintenance during glycolysis, 77–78; and hydrogen shuttles, 79–83
Regulatory enzymes: versus nonregulatory, 62–65; in control of glycolysis, 67–77
Renal chloride cells, 138
Respiratory rate: of abyssal fish, 101–103
Rosettes (glycogen), 3, 127
RQ: in hypoxic fish, 106

Sarcoplasmic reticulum (SR), 132–133
Scholander, P. F., 155
Seal: LDH levels in, 134, 136, 154; impact of diving on, 151–156; glutamine in, 159; impact of recovery on, 160–162
Sepia, 50, 51–53
Shoubridge, E., 107
Slime mold, 74
South American lungfish, 5. *See also Lepidosiren*
Sperm: hydrogen shuttle in, 81–82
Squid mantle muscle, 83–85, 87; ODH in, 44–48, 49; hydrogen shuttle in, 80
Storage glycogen, 2–5
Substrate binding, 62–65, 67–70; effect of temperature on, 89
Substrate cycling: in glycogen metabolism, 70–73; in F6P-FBP interconversion, 141, 157
Succinate, 8–9, 112–113; in helminths, 15, 18, 19, 42; in bivalves, 27, 29; in oyster heart, 30–33
Symplectoteuthis, 50
Synbranchus heart, 132–137
Szidon, J. P., 136, 142

Taegtmeyer, H., 95
Temperature: effect on oxygen uptake, 101–102; effect on FBPase, 157–158
Tilapia, 112
Trematodes, 16
Tropical fish, 117
Tuna red muscle, 95–96, 126, 129–131
Tuna white muscle, 3–4, 60–61, 126, 127; as model system, 85–95; MDH control in, 96–99

Urea cycle, 112
Uridine diphosphate glucose (UDPG), 72

Van den Thillart, G. E. E. J. M., 106, 107
van Tol, A., 157
Vasoconstriction, 146
Ventilation rate: in anoxic fish, 104
Vertebrates, 60–62; compared to cephalopods, 57–58; effect of pressure on, 102–103; brain enzyme adjustments in, 121–124. *See also* Fish; Mammals

V_{max}, 19, 63–69; for forward *vs.* reverse reactions, 160
von Brand, T., 16

Water-breathing fish, 137–138, 141
Weddell seal, *see* Seal
Whereat, A. F., 109
White muscle: *Arapaima*, 124–127; lungfish, 127–132. *See also* Tuna white muscle